Guidelines for Performing Foundation Investigations for Miscellaneous Structures

U.S. DEPARTMENT OF THE INTERIOR
BUREAU OF RECLAMATION

Technical Service Center
Denver, Colorado
2004

United States Department of the Interior
Bureau of Reclamation

MISSION STATEMENTS

The mission of the Department of the Interior is to protect and provide access to our Nation's natural and cultural heritage and honor our trust responsibilities to tribes.

The mission of the Bureau of Reclamation is to manage, develop, and protect water and related resources in an environmentally and economically sound manner in the interest of the American public.

U.S. DEPARTMENT OF THE INTERIOR
BUREAU OF RECLAMATION

Acknowledgements

Michael Szygielski and Jeffrey Farrar of the Earth Sciences and Research Laboratory, Technical Service Center, Denver, Colorado wrote this manual. Numerous sources of information were used, including sources from the Bureau of Reclamation and authoritative geotechnical engineering text books. Allen Kiene and Robert Davis of the Geotechnical Services Group performed a technical review of this manual, along with Richard Fuerst and Stephen Robertson of the Water Conveyance Group. The editor was Lelon A. Lewis. We hope that this document will be useful to all area offices, construction offices, regional geology and engineering offices, and geologists and geotechnical engineers in the Bureau of Reclamation, as well as geotechnical engineers worldwide.

Contents

Tables

Figures

Scope

This manual provides guidance on geotechnical/geological investigation requirements for miscellaneous nonhydraulic and hydraulic structures. The Bureau of Reclamation (Reclamation) has developed numerous guidance documents for design and construction of large water conveyance facilities, such as dams, pumping plants, powerplants, canals, and pipelines. These documents include the *Earth Manual*, Part 1, *Earth Manual*, Part 2, *Engineering Geology Field Manual*, *Engineering Geology Office Manual*, *Ground Water Manual*, *Drainage Manual*, *Design of Small Dams*, and *Design of Small Canal Structures* [1, 2, 3, 4, 5, 6, 7, 8]. These manuals contain important information on the content of investigations for such structures. This manual will use small examples from those larger design manuals, but the reader is encouraged to consult these manuals for more information.

The mission of the Bureau of Reclamation has changed from water resources development to water resources management. Many of the agency's projects are repairs of either existing structures, or small conveyance facilities. For these small projects, funds for investigations are limited. If investigations indicate a risk of project failure due to inadequate investigation, clients must either provide more construction funding or assume the risk. If there is inadequate investigation, the designer will make conservative design assumptions to mitigate the risk of failure. If additional investigations are required, the designer should seek more funding and make efficient use of additional funds.

Introduction

Miscellaneous hydraulic structures covered in this manual include, but are not limited to, small pipelines (no wider than 120 inches), canals (with capacities no greater than 1,000 ft³/s), check structures, diversion structures, pumping plants, storage tanks, and support buildings. This manual does not cover dam design or design for large plant structures that require more intensive investigations and laboratory testing.

This manual has four chapters. In chapter 1, *Foundation Considerations for Structures*, types of structures are defined and general investigation requirements are discussed. Chapter 2: *Background Study*, will review sources of information that can be obtained and used in the office to investigate the site without a large expenditure of funds. Chapter 3, *Site Investigation*, discusses the parameters for design of the structure's foundation and investigation methods used to obtain these parameters. Chapter 4, *Problem Soils—Soil Problems*, will address soils that in the past have been unsuitable for foundations and some soil problems that have occurred for various structures. The appendix lists *Approximate Material Characteristics*.

Since much of the material needed for a foundation investigation manual is presented in detail in various other manuals, only a brief review of these topics is given, and those publications are referenced.

The tables and charts shown in this manual are considered guides and should be used with caution. In most cases, interpretation of the data shown on the charts requires considerable geotechnical expertise and experience. When known, charts' limitations are discussed.

For new structures, foundation investigations are usually required. A multidisciplinary team is required to determine the investigation plan. In most cases, this team will consist of the structure designer, a geologist, and a geotechnical engineer. This team should meet and first look at the requirements of the structure and contents. On many projects , Reclamation is adding to existing facilities. In these cases, the explorations from the existing facilities may significantly reduce new exploration requirements. Prior to deciding the investigation requirements, the geologist and engineer should then study the site and accumulate available data.

In any study of a foundation, the first objective is to discriminate between sound and unsound foundations, and in practice to classify a foundation as adequate, inadequate, or questionable. At first, most cases will seem to fall in the questionable category, but with increased experience, these will decrease. To supplement judgment, test procedures have been developed for evaluating foundation and conditions. These may vary from simple index property tests (unit weight, water content, soil classification [Atterberg limits, grain size distribution], specific gravity, and void ratio) and visual observations to more elaborate sampling and laboratory and field tests. The extent of exploration and testing may depend on the importance of the structure, the seriousness of the soil conditions, and the steps that may be necessary to solve the problem.

The geotechnical engineer assigned the task of foundation design will have to determine the engineering properties of the materials.

The three primary engineering properties are permeability (seepage and drainage), compressibility (indexes for deformation, and for total and differential settlement), and strength (parameters for bearing capacity; stress-strain modulus, shear modulus, Poisson's ratio, cohesion, and angle of internal friction). The investigation plan will be formulated to allow the geotechnical engineer to estimate or directly measure these properties.

Chapter 1

Foundation Considerations for Structures

In the exploration of any foundation, it is of little value to examine only the exposed surface, because the subsoil strength at depth determines footing requirements. Rules for reasonable depths of investigations are based on the theoretical distribution and extent of influence of pressures under a foundation.

Investigations for miscellaneous structures can be classified into two groups, those for point structures and those for line structures. More detail on the discussions presented below can be found in the *Earth Manual* [1]. A first examination of a project consists of a review of the tentative plans for structures, including the purpose, size, location, typical loadings, value of water, and any unusual features.

1.1 Point Structures

For structures such as small buildings, small pumping plants, transmission towers, and bridge piers, a single test hole is often adequate. Larger structures may require more test holes. When the exact location of a structure depends on foundation conditions, the number of test holes required should be increased. Two or three test holes are used for preliminary exploration to establish general foundation conditions; the investigation requirement can usually be reduced for later stages.

Small structures often do not apply appreciable loads to foundations, yet soil or rock conditions must be identified during investigations. Generally these sites can be investigated with drill holes or test pits, depending on the intended foundation elevation. The standard penetration resistance test (SPT) can be used to evaluate estimated bearing capacity for these structures.

Open-air pumping plant pads and manifold structures founded on single slabs are often very lightly loaded. If founded on dry soils, and later leakage causes some loss of support to the slab, slabs can be improved by mud jacking. Differential settlement is always a concern, even with smaller structures. Pumping plant manifolds can generate considerable lateral loads and vibrations, which must be evaluated.

Heavier structures with loads greater than 2 to 3 t/ft^2, may require more detailed analysis, including sampling and laboratory consolidation testing.

Pumping plants should be given special attention because of the intense influence of vibrations and sensitivity to settlement. Continual economic operation of pumps is possible only if the settlement of the foundation is reduced to an absolute minimum.

Generally, holes are drilled at the approximate location of the structure corners. Additional holes should be bored at the center location and at the location of any heavy bearing wall.

In rock with apparently adequate bearing characteristics, the bore holes need to penetrate only 25 feet, providing it is reasonably certain softer materials do not lie below this depth.

1.2 Line Structures

Exploration requirements for the foundations of canals, pipelines, and roads vary considerably according to the size and importance of the structure and according to the character of the ground through which the line structure is to be located. Spacing of holes or other explorations will vary, depending on the need to identify changes in subsurface conditions. Where such structures are to be located on comparatively level ground with uniform soils such as the plains areas, fewer holes along the alignment may suffice for foundation investigation requirements. In certain instances, special investigations may be required, such as in-place density measurements, for pipelines or hand cut block samples to study collapse potential in areas of low density soils.

Line structure investigations should begin by studying the geology of the proposed alignments. These studies should include use of available geologic data as outlined in detail in the *Earth Manual* [1]. The exploration geologist should survey the complete alignment prior to developing an intrusive exploration program. This is because the locations of investigations could easily identify problem areas such as rock outcrops or swamp deposits. The geologist should also develop a surface geological map for the alignment, to identify the various deposits to be encountered and characterized.

Due to the length of line structures, investigations must be located judiciously. Often, the locations of borings and test pits will be placed in critical areas. In other areas, consideration should be given to use of faster and cheaper methods, such as disturbed power augering, to confirm anticipated geologic conditions in between more detailed locations.

For line structures, limits of rock excavation, if encountered, must be identified.

1.2.1 Canals

Investigations for canals, laterals, and sublaterals should consider the foundation materials encountered, cut slope stability, cut and fill quantities, and any special borrow requirements for materials such as lining and drains. Investigations for canals should consider that wetting of the canal prism will occur, and therefore identification of collapsible or expansive soils must be carefully considered. The typical concrete-lined canal used by Reclamation, cannot accommodate excessive earth deformation without unacceptable cracking and increased leakage. In cases where deformations may be unacceptable, alternatives exist, such as the use of earth linings or use of geomembrane linings with protective covers of gravel or shotcrete.

Investigations for canals normally must consider cut slope stability and the need for stability berms, maintenance roads, and control of runoff. Canal cuts in clays can be problematic if the cut causes instability or if the clays soften and lose strength. Cuts in stiff, fissured clays, expansive clays, and clay shales have been known to initiate failures. In these cases, often just a thin, weak seam can be the culprit.

Slope failures in clays can also be accentuated by long term softening and by rapid drawdown conditions during canal operations. Some clay

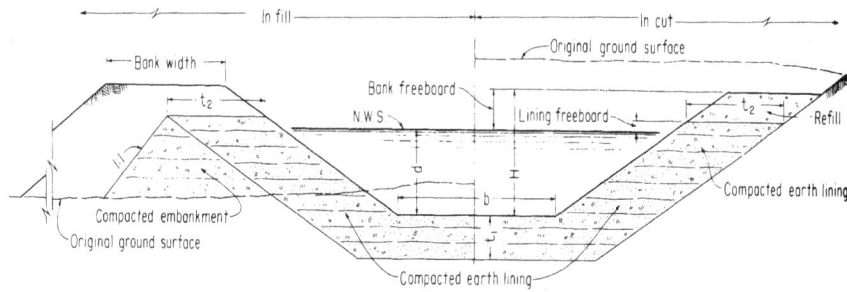

Figure 1.—Typical compacted-earth-lined canal section.

soils soften and swell when wetted and weaken to the extent that much flatter slopes fail. Canal side slopes of marginal stability should never be subjected to rapid drawdown during dewatering operations, as side slope sloughing is likely. A rule-of-thumb procedure when dewatering canals is to lower the water level not more than a foot a day to permit the soil-water pressures within the soil to dissipate. Unstable clays, as in the Friant Kern Canal, have required lime treatment for stabilization [9].

In fill sections, impervious soils are required for water retention in embankment sections, and in some cases, filter and drain materials may be required. Dispersive soils tests are routinely performed in canal investigations. Dispersive soils should be avoided in fill sections, or where natural seepage exit conditions can allow for piping failures. Dispersive or erosive soils can cause loss of soil behind a concrete-lined canal. Dispersive soils can be used, but must be identified and treated with lime or protected with filters to prevent failures.

Linings are most always used in canal sections to reduce seepage losses (fig. 1) [10]. Investigations should consider any special material requirements for lining materials. Canal side slopes can be protected by a gravel beach belt or a moderately plastic, compacted clay lining. Thin compacted clay linings are difficult to construct

by rolling on canal slopes, and Reclamation uses thicker equipment width linings or alternate materials such as geomembranes. Clay lining materials which are sufficiently erosion resistant should have sufficient plasticity, as shown on figure 2. If dispersive clays are present or erosion is reoccurring or occurring in turbulence zones, the use of gravel or rock protective covers is warranted.

The occurrence of seeps and ground water seepage against thin concrete linings could require the use of drains and weep holes to relieve pressures behind the lining. Explorations for canals typically consist of combinations of test pits, drill holes, and power auger holes.

1.2.2 Pipelines

Pipelines have some special investigation requirements, depending on the line length, size, and type of pipes being used. Pipe can be classified as "rigid" or "flexible" depending on the type and size of pipe. Generally, for pipe less than 10 inches in diameter, it is not economical to provide compacted embedment support, and backfill may be dumped and tamped. Figure 3 shows Reclamation's standard pipe installation drawing for pipe with select embedment.

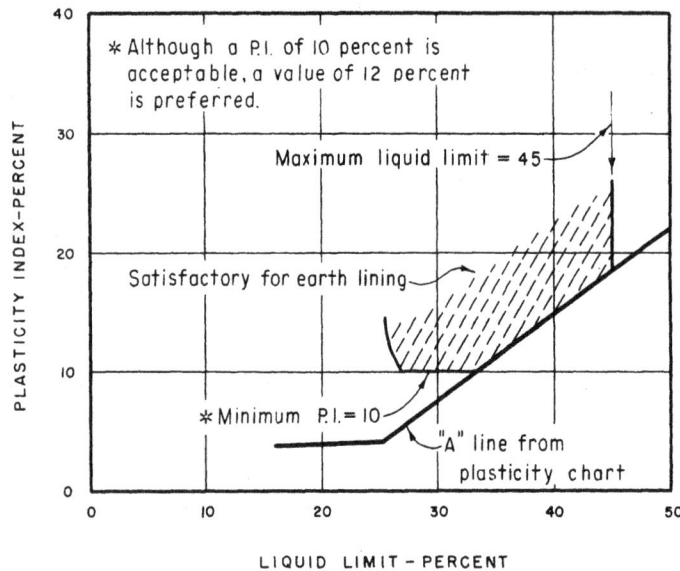

Figure 2.—Plasticity criterion for impervious, erosion-resistant compacted earth linings.

Pipe larger than 12 inches (polyvinyl chloride [PVC], high density polyethylene [HDPE], pretensioned concrete cylinder, steel, and fiberglass) and ductile iron larger than 24 inches in diameter are classified as flexible pipe.

Flexible pipe design is based on allowable deflections, generally less than 3 to 5 percent. Flexible pipe deflection is governed by the stiffness of the trench wall material and backfill through use of an empirical modulus of soil reaction, $E\yen$ (table 1 [11, 12]). Trench width is governed by springline support provided by the in-place soils, with weaker soils requiring larger trenches with more compacted backfill for the pipe. Pipe investigations should determine the estimated $E\yen$ conditions of the natural soil. This is accomplished by test pitting, performing in-place density testing, and determining the degree of compaction or relative density of the soils by laboratory compaction test. Natural $E\yen$ may also be estimated by SPT blow count or compression tests (table 2 [11]). The modulus

of soil reaction of the in-place soils can also be estimated by penetration resistance testing. For short pipelines, the designer may opt for increased wall thickness of the pipe to avoid costly investigations. An economic determination of investigation cost versus cost of stronger pipe must be estimated.

Rigid pipe depends on adequate bearing capacity at the invert. This is because most of the load is transmitted at the base through a bedding angle of 20 to 30 degrees. For rigid pipe, it is necessary to identify weak soils, which could require overexcavation and replacement.

For pipelines anticipated to encounter expansive soils, flexible pipe is desired, because of possible large stress concentrations at the bottom of rigid pipe. Alternative treatments for expansive soils include overexcavation and lime treatment, or overexcavation and backfill with a coarse rock layer.

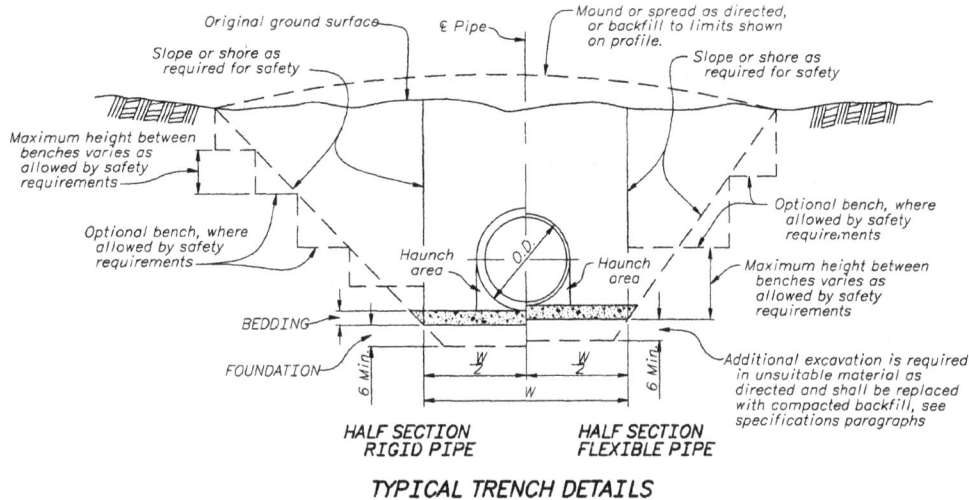

TYPICAL TRENCH DETAILS

MINIMUM INSTALLATION WIDTH

PIPE I.D. (INCHES)	W (FEET)
6 and less	2.0
Over 6 thru 18	$\frac{1}{12}$ (O.D. + 20)
Over 18	$\frac{1}{12}$ (O.D. + 36)

GRADATION LIMITS FOR SELECT MATERIAL

SIZE *	PERCENT BY WEIGHT
Passing No. 200 sieve	5 or less
Passing No. 50 sieve	25 or less

* Maximum size shall not exceed 3/4 inch.

SIDE CLEARANCE TABLE

TRENCH TYPE	MINIMUM SIDE CLEARANCE [W1] (INCHES)
1	10 INCHES FOR 12" THRU 18" I.D. 18 INCHES FOR OVER 18" I.D.
2	ONE O.D.
3	TWO O.D.

For location of Trench Types, see Specifications.

Figure 3.—Reclamation's typical trench for pipe with select embedment.

One cannot assume that the pipeline will not leak, and collapsible soils could be detrimental to pipe support. Therefore, potentially collapsible soils must be identified.

Typical pipe backfill embedment is either clean, cohesionless soil with less than 5 percent fines, or soil cement slurry. The soil cement slurry is also known as "flowable fill" or "controlled low strength material" [12]. Therefore, investigations must also determine the source of these materials. Recently, a trend toward use of soil cement slurry has increased. For soil cement slurry backfill, the contractor is allowed to use sands with less than 30 percent silty fines. He may elect to use these soils if they are present in the excavation. Clayey soils have

been allowed on non-Reclamation projects, but the long term stability is a concern, and Reclamation does not typically allow these soils for soil cement slurry.

Backfill above the pipeline must be moderately compacted to prevent subsequent settlement. Higher levels of compaction are required in critical areas, such as road crossings. The investigations should contain compaction test data on the soils to be compacted, especially the need for water to facilitate compaction.

Pipeline that are constructed with cohesionless sand/gravel backfill will act as French drains after construction. The pipe alignment could affect shallow ground water levels and drainage

Table 1.—Selection of trench type for compacted embedment with E´ = 3000

Trench wall soil classification Unified Soil Classification System	Degree of compacton* of trench walls		
	Slight	Moderate	High
	< 85% P < 40% RD	≥ 85 to < 95 % P ≥ 40 to < 70% RD	≥ 95% P ≥ 70% RD
Highly compressible fine-grained soils CH, MH, OH, DL Peat, swamps, bogs	(trench wall E´ < < 100) Trench type 3		
Fine-grained soils Soils with medium to no plasticity and with less than 30% coarse grained particles CL, ML (or CL-ML, CL/ML, ML/CL)	(trench wall E´ = 200) Trench type 3	(trench wall E´ = 400) Trench type 3	(trench wall E´ = 1500) Trench type 2
Sandy or gravelly fine-grained soils Soils with medium to no plasticity and with 30% or more coarse-grained particles CL, ML (or CL-ML, CL/ML, ML/CL) **Coarse-grained soils with fines** Sands, gravels with more than 12% fines GC, GM, SC, SM (or any soil beginning with one of these symbols (i.e., SC/CL))	(trench wall E´ = 200) Trench type 3	(trench wall E´ = 1000) Trench type 2	(trench wall E´ = 2500) Trench type 1
Clean coarse-grained soils Sands, gravels with 12% or less fines GW, GP, SW, SP or any soil beginning with one of these symbols (i.e., GP-GM)	(trench wall E´ = 700) Trench type 3	(trench wail E´ = 2000) Trench type 2	(trench wall E´ = 3000) Trench type 1
Rock, sandstone, shale Highly cemented soils, etc.	(trench wall E´ > > 3000 Trench type 1		

* % P = % compaction
RD = % relative density; E´ = shown as lb/in²

conditions in these areas. Investigations should consider the impacts of drainage on the surrounding land under this condition. If impacts are adverse, such as wetland drainage, the backfill can be specified to contain impervious plugs to prevent the French drain effect.

1.3 Roadways

Investigations for roadways are very similar to those for canal structures. Considerations include foundation conditions to be encountered, cut slope stability, cut and fill quantities, and special borrow needs. In fill areas, the foundation needs to be evaluated to

Table 2.—Values of $E_n{}'$ for native, in situ soils [11]

Granular		Cohesive		
Blows/ft [1]	Description	q_u (t/ft^2) [2]	Description	$E_n{}'$ (lb/in^2)
> 0 - 1	very, very loose	> 0 - 0.125	very, very soft	50
1 - 2	very loose	0.125 - 0.25	very soft	200
2 - 4	very loose	0.25 - 0.50	soft	700
4 - 8	loose	0.50 - 1.00	medium	1,500
8 - 15	slightly compact	1.00 - 2.00	stiff	3,000
15 - 30	compact	2.00 - 4.00	very stiff	5,000
30 - 50	dense	4.00 - 6.00	hard	10,000
> 50	very dense	> 6.00	very hard	20,000
Rock	-----	--------	----------	> 50,000

[1] Standard penetration test per ASTM D 1586
[2] Data can be obtained from ASTM unconfined compression tests

allow determination of embankment settlements. Fills on cross slopes should be carefully evaluated.

1.4 Backfill Materials

Sources of construction materials are obtained from required excavation, adjacent borrow, and distant borrow. In nearly all cases, the material from required excavation will be used somewhere as backfill. Additional borrow material is obtained from areas adjacent to the structure, and test holes are not required if alignment test holes are sufficiently close to ensure the availability of good materials. If readily available materials pose undesirable characteristics, it may be necessary to investigate distant borrow to obtain materials for blending.

Select free draining sand gravel materials are often required for backfill about pumping

plants, filters and drains, and pipeline construction. If these soils are not available from required excavation, borrow sources will need to be identified. Local concrete aggregate plants are an excellent source for free draining materials. Concrete sand meeting requirements of ASTM C-33 is an excellent filter material and can filter most all fine grained soils from piping and internal erosion. The Materials Engineering and Research Laboratory of Reclamation's Technical Service Center, Denver maintains quarry records on concrete aggregates and is a good place to start to look for these soils.

Extensive borrow investigation for small structures should not be required. For most cases, simply identifying the soil classification should be enough. In some cases, it's necessary to estimate shrinkage and swell from potential borrow areas. A list of shrinkage and swell factors for a wide variety of materials is given in the appendix.

Chapter 2

Background Study

Topography is usually required, even for small construction projects. The resolution required depends on the size of the structure. Topography data should be acquired early in the project to aid in planning investigations and performing conceptual designs. For more information on topography, consult the *Engineering Geology Field Manual* [3] and the *Earth Manual* [1, 2].

The team must have at least a general knowledge of the foundation and material requirements for the various facilities under consideration, if the investigations are to be accomplished effectively and efficiently.

The team should review local conditions, features, and similar construction in the area.

The background study should lead to an appraisal of the general surface and subsurface conditions, as well as an evaluation of the foundation conditions of alternate sites, if necessary.

2.1 Soil Characterization

To select soils, it is first necessary to identify and classify them according to a system, which is related to their physical or engineering properties. Reclamation has adopted the Unified Soil Classification System (USCS); see designations USBR 5000 and 5005, *Earth Manual* [2]. The system provides for both a visual method and a method based on laboratory tests. The proportions of the soil components (gravel, sand, silt, or clay) and the plasticity (the stickiness or cohesiveness) of the silt and clay fraction are defined. The visual method uses hand tests and visual observations and is not expected to be as precise as the laboratory method. With a limited amount of training and an interest in soil classification, one can identify soils with sufficient accuracy to classify them according to the 15 basic soil groups, as shown on tables 3 and 4 [2, USBR 5000].

Agricultural soil surveys are a good starting point to investigate the soils present at a site. The U.S. Department of Agriculture (USDA) soil triangle of the basic soil texture classes is shown on figure 4 [1, p. 71]. USDA National Resources Conservation Service soil survey data for surficial soils are readily available for almost all of the United States. Most of these soils surveys also report soil type using the Unified Classification System.

The next step in the selection of soils for use in construction is to relate the group or type of soil and its engineering properties (strength, permeability, and compressibility). The "Engineering use chart" shown on table 4 [1, p. 51], has done this broadly. This chart may be used as a guide for evaluating the relative desirability of the soil types for various uses and estimating their important properties.

Table 3.—Soil classification chart—laboratory method

Criteria for assigning group symbols and group names using laboratory tests [1]				Soil classification	
				Group symbol	Group name [2]
Coarse-grained soils — More than 50% retained on No. 200 sieve	Gravels — More than 50% of coarse fraction retained on No. 4 sieve	Clean gravels — Less than 5% fines [3]	$Cu \geq 4$ and $1 \leq Cc \leq 3$ [\5]	GW	Well-graded gravel [6]
			$Cu < 4$ and/or $1 > Cc > 3$ [\5]	GP	Poorly graded gravel [6]
		Gravels with fines — More than 12% fines [3]	Fines classify as ML or MH	GM	Silty gravel [6,7,8]
			Fines classify as CL or CH	GC	Clayey gravel [6,7,8]
	Sands — 50% or more of coarse fraction passes No. 4 sieve	Clean sands — Less than 5% fines [4]	$Cu \geq 6$ and $1 \leq Cc \leq 3$ [\5]	SW	Well-graded sand
			$Cu < 6$ and/or $1 > Cc > 3$ [\5]	SP	Poorly graded sand [9]
		Sands with fines — More than 12% fines [4]	Fines classify as ML or MH	SM	Silty Sand [7,8,9]
			Fines classify as CL or CH	SC	Clayey sand [7,8,9]
Fine-grained soils — 50% or more passes the No. 200 sieve	Silts and clays — Liquid limit less than 50	inorganic	PI > 7 and plots on or above "A" line [10]	CL	Lean clay [11,12,13]
			PI < 4 or plots below "A" line [10]	ML	Silt [11,12,13]
		organic	$\dfrac{\text{Liquid limit-oven dried}}{\text{Liquid limit-not dried}} < 0.75$	OL	Organic clay [11,12,13,14] Organic silt [11,12,13,15]
	Silts and clays — Liquid limit 50 or more	inorganic	PI plots on or above "A" line	CH	Fat clay
			PI plots below "A" line	MH	Elastic silt [11,12,13]
		organic	$\dfrac{\text{Liquid limit-oven dried}}{\text{Liquid limit-not dried}} < 0.75$	OH	Organic clay [11,12,14,15] Organic silt [11,12,14,16]
Highly organic soils		Primarily organic matter, dark in color, and organic odor		PT	Peat

[1] Based on the material passing the 3-in (75-mm) sieve.
[2] If field sample contained cobbles and/or boulders, add "with cobbles and/or boulders" to group name.
[3] Gravels with 5 to 12% fines require dual symbols
　　GW-GM well-graded gravel with silt
　　GW-GC well-graded gravel with clay
　　GP-GM poorly graded gravel with silt
　　GP-GC poorly graded gravel with clay
[4] Sands with 5 to 12% fines require dual symbols
　　SW-SM well-graded sand with silt
　　SW-SC well-graded sand with clay
　　SP-SM poorly graded sand with silt
　　SP-SC poorly graded sand with clay

[5] $Cu = D60/D10$　　$C_c = \dfrac{(D_{30})^2}{D_{10} \times D_{60}}$

[6] If soil contains $\geq$ 15% sand, add "with sand" to group name.

[7] If fines classify as CL-ML, use dual symbol GC-GM, SC-SM.
[8] If fines are organic, add "with organic fines" to group name.
[9] If soil contains $\geq$ 15% gravel, add "with gravel" to group name.
[10] If the liquid limit and plasticity index plot in hatched area on plasticity chart, soil is a CL-ML, silty clay.
[11] If soil contains 15 to 29% plus No. 200, add "with sand" or "with gravel" whichever is predominant.
[12] If soil contains $\geq$ 30% plus No. 200, predominantly sand, add "sandy" to group name.
[13] If soil contains $\geq$ 30% plus No. 200, predominantly gravel, add "gravelly" to group name.
[14] PI $\geq$ 4 and plots on or above "A" line.
[15] PI < 4 or plots below "A" line.
[16] PI plots on or above "A" line.
[17] PI plots below "A" line.

Table 4.—Tabulations of engineering properties of compacted soils used in earth structures [13]

Typical names of soil groups	Group symbols	Important engineering properties				Relative desirability for various uses (from 1 to 15, best to worst)										
		Permeability when compacted	Shear strength when compacted and saturated	Compressibility when compacted and saturated	Workability as a construction material	Rolled earthfill dams			Canal sections		Foundations		Fills		Foundations	Pipe embed.
						Homogeneous embankment	Core	Shell	Erosion resistance	Compacted earth lining	Seepage important	Seepage not important	Frost heave not possible	Frost heave possible	Surfacing	Support capability
Well-graded gravels, gravel-sand mixtures, little or no fines	GW	Pervious	Excellent	Negligible	Excellent	-	-	1	1	-	-	1	1	1	3	5
Poorly graded gravels, gravel-sand mixtures, little or no fines	GP	Very pervious	Good	Negligible	Good	-	-	2	2	-	-	3	3	3	-	5
Silty gravels, poorly graded gravel-sand-silt mixtures	GM	Semipervious to impervious	Good	Negligible	Good	2	4	-	4	4	1	4	4	9	5	4
Clayey gravels, poorly graded gravel-sand-clay mixtures	GC	Impervious	Good to fair	Very low	Good	1	1	-	3	1	2	6	5	5	1	6
Well-graded sands, gravelly sands, little or no fines	SW	Pervious	Excellent	Negligible	Excellent	-	-	3 if gravelly	6	-	-	2	2	2	4	3
Poorly graded sands, gravelly sands, little or no fines	SP	Pervious	Good	Very low	Fair	-	-	4 if gravelly	7 if gravelly	-	-	5	6	4	-	1
Silty sands, poorly graded sand-silt mixtures	SM	Semipervious to impervious	Good	Low	Fair	4	5	-	8 if gravelly	5, erosion critical	3	7	8	10	6	2
Clayey sands, poorly graded sand-clay mixtures	SC	Impervious	Good to fair	Low	Good	3	2	-	5	2	4	8	7.	6	2	3.
Inorganic silts and very fine sands, rock flour, silty or clayey fine sands with slight plasticity	ML	Semipervious to impervious	Fair	Medium	Fair	6	6	-	6	6, erosion critical	6	9	10	11	-	3
Inorganic clays of low to medium plasticity, gravelly clays, sandy clays, silty clays, lean clays	CL	Impervious	Fair	Medium	Good to fair	5	3	-	9	3	5	10	9	7	7	6

Table 4.–Tabulations of engineering properties of compacted soils used in earth structures [13]

Typical names of soil groups	Group symbols	Important engineering properties				Relative desirability for various uses (from 1 to 15, best to worst)										
		Permeability when compacted	Shear strength when compacted and saturated	Compressibility when compacted and saturated	Workability as a construction material	Rolled earthfill dams			Canal sections		Foundations		Fills		Surfacing	Pipe embed.
						Homo-geneous embank-ment	Core	Shell	Erosion resistance	Compacted earth lining	Seepage important	Seepage not important	Frost heave not possible	Frost heave possible		Support capability
Organic silts and organic silt-clays of low plasticity	OL	Semipervious to impervious	Poor	Medium	Fair	8	8	-	-	7, erosion critical	7	11	11	12	-	10
Inorganic silts, micaceous or diatomaceous fine sandy or silty soils, elastic silts	MH	Semipervious to impervious	Fair to poor	High	Poor	9	9	-	-	-	8	12	12	13	-	9
Inorganic clays of high plasticity, fat clays	CH	Impervious	Poor	High	Poor	7	7	-	10	8, volume change critical	9	13	13	8	-	9
Organic clays of medium to high plasticity	OH	Impervious	Poor	High	Poor	10	10	-	-	-	10	14	14	14	-	10
Peat and other highly organic soils	Pt	-	-	-	-	-	-	-	-	-	-	-	-	-	-	10

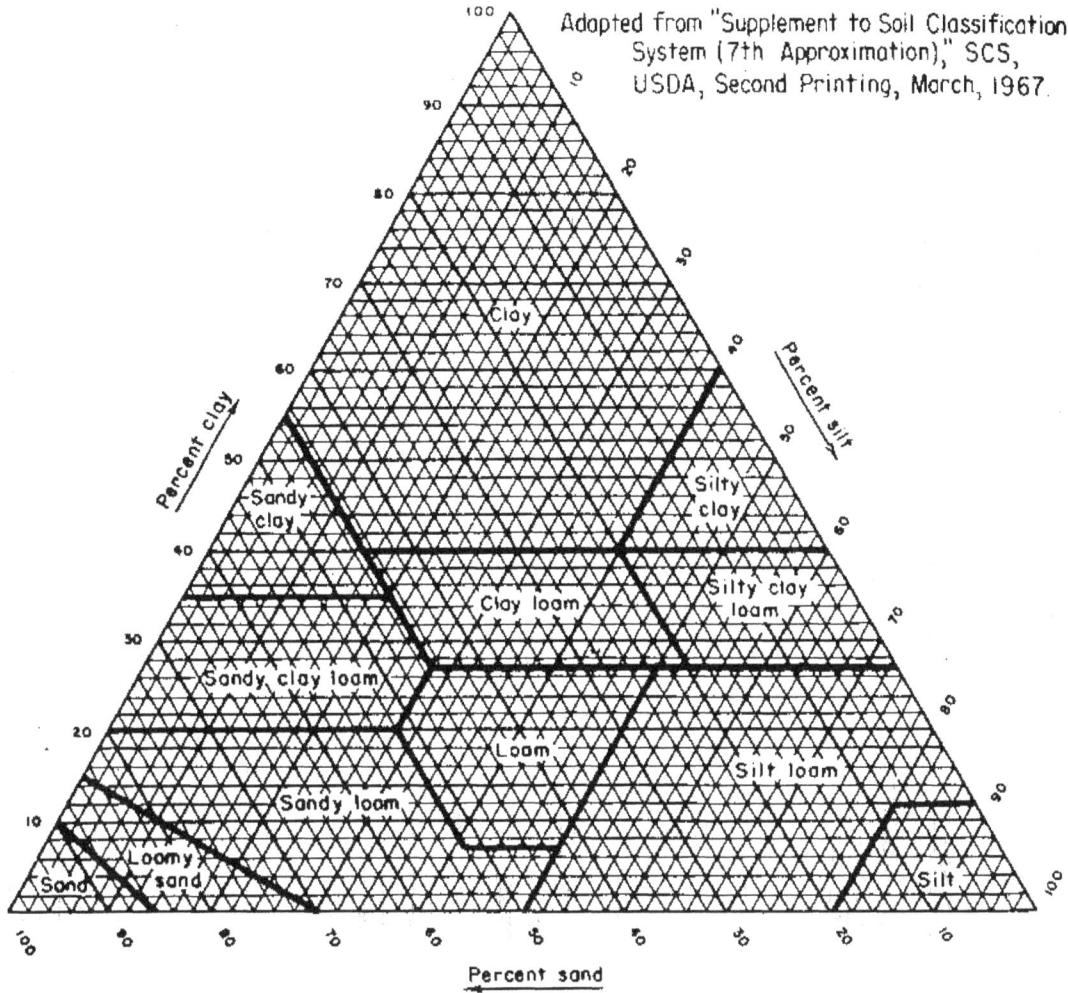

Adapted from "Supplement to Soil Classification System (7th Approximation)," SCS, USDA, Second Printing, March, 1967.

* Very fine sand (0.05 – 0.1) is treated as silt for family groupings; coarse fragments are considered the equivalent of coarse sand in the boundary between the silty and loamy classes.

COMPARISON OF PARTICLE-SIZE SCALES

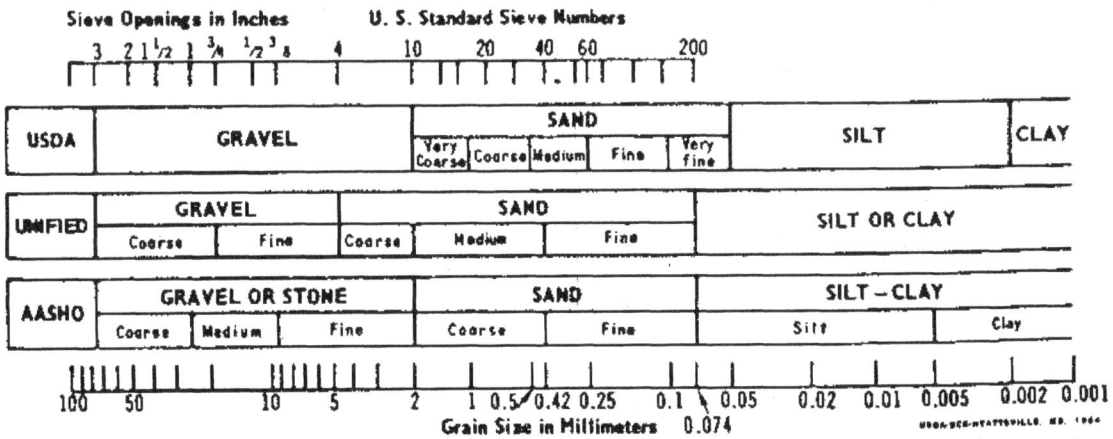

Figure 4.—Soil triangle of the basic soil texture classes (Natural Resources Conservation Service).

Operation and maintenance (O&M) work in water systems is concerned with the repair or replacement of completed structures. Failures are usually due to the action of water. Insofar as earthwork is concerned, an entire structure such as a canal bank or highway fill may wash out or slide, or the soil backfill around a buried structure may erode or pipe, causing partial failure. Knowledge of soils and their physical properties, particularly as influenced by water, will aid in analyzing the cause of failure and selecting the most suitable soils for repair or the proper method of processing and replacement.

2.2 Maps and Photo Information

Map and photo information from the U.S. Geologic Survey (USGS) are discussed in the *Earth Manual* [1] starting on page 64. The sources discussed are:

- Geologic maps
- Hydrologic maps
- Geophysical maps
- Topographic maps
- Hazard maps
- Agricultural
- Soil Maps
- Remote Sensing Techniques

Topography is required for most all construction projects. USGS public domain topography with 20-foot contours is not sufficient for construction, where 1- to 5-foot contours are required. However, it may be useful in initial background studies. Detailed topography should be obtained prior to construction. Local topography contractors can be contacted at the beginning of project for detailed topography. Construction topography must include benchmarks, locations of manmade structures, and all utilities.

2.3 Surface Land Form Types

Surface land form types are discussed in the *Earth Manual* [1] starting on page 77. Table 5 summarizes the landform descriptions and gives some broad generalizations about the engineering characteristics and applications of these soil types associated with their particular land forms. Table 5 lists possible foundation problems typically associated with the land form types.

2.3.1 Surface Water

Existing surface water, stream flow, and runoff can be determined from topographic maps. Stream flow data may be available from gauging stations operated by the U.S. Geological Survey or other State and local agencies. Seasonal flucuations of water courses must be determined. If construction alignment crosses water courses, investigations may be required for dewatering.

2.4 Subsurface Considerations

Subsurface geotechnical exploration is performed primarily for three purposes: (1) to determine what distinct masses of soil and rock exist in a foundation or borrow area within the area of interest, (2) to determine the dimensions of these bodies, and (3) to determine their engineering properties [1, p. 85].

In the exploration of any foundation, it is of little value to examine only the exposed surface of the site alone, because the subsoil strength at

Table 5.—Possible foundation problems typically associated with land form types

Landform	Possible foundation problems or other problems
Alluvial deposits	
Stream Channel	Loose liquefiable soils
Flood Plain	Soft compressible soils
	Loose liquefiable soils
Terrace	Cemented soils
Alluvial Fan	Well sorted soils good for borrow
Slope wash or colluvial deposits	Loose, collapsible Soils
Lacustrine deposits	Soft compressible soils
Glacial deposits	
Tills or glacial	Nonuniform soils and settlement
	Dense soils
Outwash or glaciofluvial	Nonuniform soils and settlement
	Loose, liquefiable soils
Eolian deposits	
Loess	Collapsible low density soils
Dune	Loose, low density sands
	Liquefaction of saturated
Residual soils	Nonuniform weathering of parent rock and subsequent settlement
Shales	Cut slope stability
	Compaction of excavated shale is difficult
	Friable shales and foundation deterioration
	Expansive clay
Desert soils	Caliche hardpan
Vocanic tuff	Variable rock quality
	Poor rock quality for aggregates
Volcanic ash	Alters to clays
Brecciated (sheared) rock	Poor quality rock aggregates
	Clay gouge in shear zones

depth determines footing requirements. There are rules for reasonable depths of investigations based on the theoretical distribution and extent of influence of pressures under a foundation. Sampling depth for various foundations are discussed in section 3.2.4, *Sampling Depth*.

2.4.1 Soils

In a foundation, designers are interested in knowing the condition of the soil or rock. It is, of course, of value to know the type of soil or rock, such as clay, silt, sand or gravel, or shale, which could be determined by accessible sampling or sampling from boreholes. But unless the samples permit us to interpret the firmness and denseness of the in-place soil, designers cannot fully determine the foundation supporting capacity. Many direct and indirect methods can determine foundation conditions.

Table 6 lists the parameters needed to define the subsurface engineering properties for permeability, compressibility, and strength. When reviewing background information, the availability of these parameters should be noted. The most important parameter affecting engineering properties is the consistency or degree of compaction, which is a combination of in-place density compared to laboratory maximum.

2.4.2 Ground Water

Ground water has a major influence on foundation performance. Fluctuations in ground water levels should be anticipated; a significant rise in water levels could compromise foundation bearing capacity. Dewatering of foundations requires information on the soils present and sources of ground water. Ground water contamination should be identified if it is suspect. Ground water quality and the soils present affect corrosion potential. Water quality tests, primarily for sulfates and chlorides, of ground water and soils in contact with structures should be evaluated. Water quality information may be available from existing wells in the area. Sulfates adversely affect concrete, and high sulfates should be identified.

Table 6.—Parameters needed to define the subsurface engineering properties for permeability, compressibility, and strength

Engineering properties	Permeability	Compressibility	Strength
Classification Gradation Atterberg limits	x	x	x
Specific gravity	x	x	x
Natural moisture/density	x	x	x
Compacted moisture/density relationship	x	x	x
Degree of compaction	x	x	x
Coefficient of permeability	x	x	x
Coefficient of consolidation		x	
Angle of internal friction			x
Cohesion			x

2.5 Existing Structures

2.5.1 Adjacent Structures

Generally, foundation investigations will be either confirmatory or exploratory. Where a structure is to be placed near an existing structure considerable data are usually available with regard to adjacent structural behavior, types of foundation, and subsurface conditions. Investigations should be planned to correlate with and extend existing information. Essentially, the investigations under these conditions are to confirm that the soil conditions under the proposed structure are consistent with those under existing structures and thus, permit the use of observed behavior of earlier structures, in evaluating the design and performance of the new structure.

An effort should be made, wherever possible, to obtain information concerning foundation investigations, design (especially of foundations), construction experience, and behavior of structures of significant size in the area of interest. Such information may include settlement, boring logs, field tests and measurements, ground water conditions, and foundation or construction problems. The Earth Sciences and Research Laboratory may have soil testing reports of existing Reclamation structures. Local geotechnical consulting firms and drilling companies may be a good source of information for new areas.

The physical conditions and foundations of nearby structures should always be investigated, if possible. Such studies are especially important where an existing structure or facility may be affected by the proposed building. Examples include the effects of necessary excavations, additional settlement resulting from an increase of soil stresses caused by the weight of the new structure (especially at sites with deep beds of soft, compressible soil), or the effects of pile driving or dewatering. Changes in surface water drainage due to the proposed structure should be considered for both surface water runoff, and the effects on local groundwater conditions.

Buried structures such as pipes, cables, or subways pose special problems, in that frequently, their locations may be known only approximately. Actual locations of these facilities should be established to ascertain that they would not interfere with the proposed construction. The effects of the proposed construction, such as settlement caused by the weight of the structure or by dewatering or lateral distortions caused by excavations, must be considered. Pipelines carrying fluids, such as water mains or sewers, are particularly critical, as modest distortion may lead to leaks that may cause failure of cut slope or sheeting of excavations. Also note surface water runoff and the effects on local groundwater conditions.

Where disturbance of existing structures could occur, a careful survey should be made of the physical conditions of the structure, including the mapping of all cracks. Reference marks should be established for checking settlement and lateral displacement.

Existing structures in similar soil deposits may show signs of corrosion or concrete sulfate attack. These occurrences should be noted and the need for cathodic protection systems and sulfate resistant cement aniticpated.

2.5.2 Road Cuts and Fills

Existing slopes along road cuts would give insight to the possible subsurface stratification of a site. If the road cut has been there for any length of time, the slope may have stabilized and it should be noted if the slope is different from the initial design as shown on figure 5.

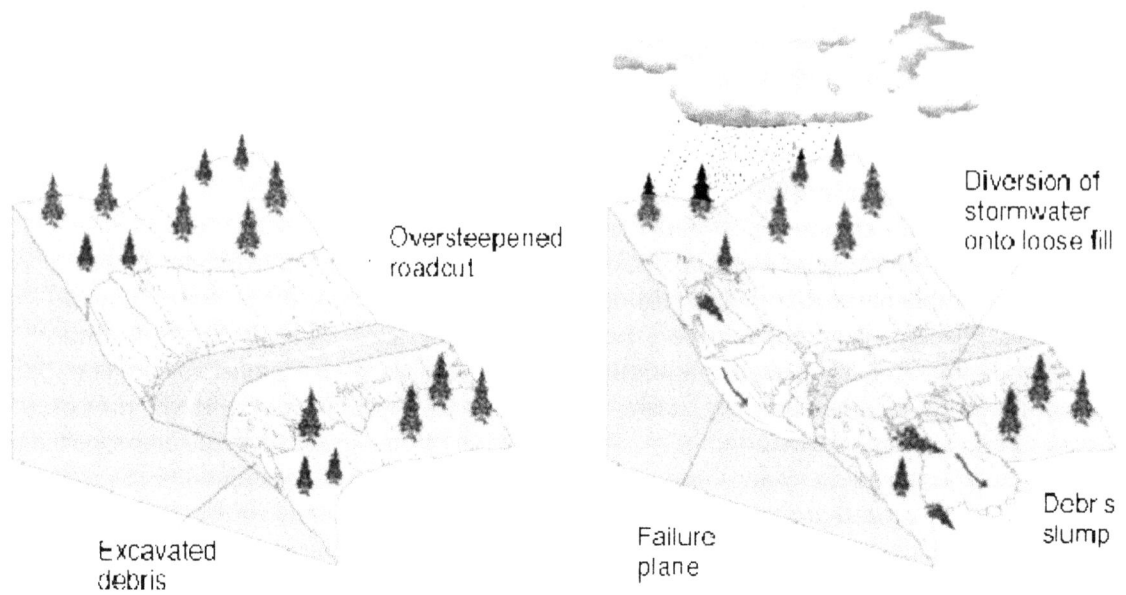

Figure 5.— Road cuts can destabilize a slope by oversteepening (left), or loading the head of a slope (right).

Chapter 3

Site Investigation

3.1 Initial Surface Site Reconnaissance

Site reconnaissance can proceed after the necessary precursory information is obtained. Geologic, topographic, and soil survey maps and reports should be reviewed. A team consisting of a geologist, geotechnical engineer, and design engineer should perform the site reconnaissance. The site should be viewed completely. The team should also inspect the surrounding area as discussed in chapter 2. For example, for cut slope stability, surrounding highway cut slopes should be inspected. Existing structures can be surveyed. The team should plan to visit with authorities familiar with construction in the project area to determine any unknown or unseen problems. During the inspection, alternate site structure alignments should be considered.

3.2 General Considerations for Specific Site Investigations

Site investigations are seldom a simple procedure requiring only conscientious adherence to a set of hard and fast rules. Unless the team is guided by mature judgment and has had a varied practical experience in this field, much time and money may be wasted. A thorough knowledge of geology of sedimentary and other unconsolidated masses is an asset of

inestimable value, because factual knowledge is always limited to soil conditions along vertical lines spaced far apart. The results of interpolation and the estimate of possible scattering can be very misleading, unless the investigators have a fairly clear conception of the soil profile under investigation. A knowledge of the geology of the region is also needed to determine whether beds beneath the structure site have ever been subjected to greater loads than at present and, if so, to provide a basis for estimating the magnitude of the additional pressure. The results of site investigations are ultimately condensed into a set of assumptions that constitute the basis for design, provided that the soil profile is relatively simple. However, if the methods for investigations are judiciously selected and intelligently used, fairly reliable limiting values can be obtained for circumstances encountered.

3.2.1 General

Subsurface investigations are costly and should only be considered after the investigation team performs initial site reconnaissance. The team should list the design issues and data requirements. After consideration of the engineering properties of needed, and with consideration of the required costs of the investigations, the team should select the proper investigation method. The remainder of this manual provides numerous references for estimating engineering properties. The team should consider the importance of the required data and the risk involved, and decide the level of investigation required to satisfy design data requirements.

Reclamation forms an investigation team that develops a formal "Field Exploration Request" (FER). This is the best estimation by the team of the work required. The FER should be flexible if methods are not working or revealed site conditions dictate changes. In addition, FER's are normally phased to allow for review of the data adequacy. Communication between field and design staff throughout the exploration process saves both time and effort. A clear understanding of the design issues being investigated allows the field personnel to adapt to potentially different geologic conditions than expected.

3.2.2 Surface Water

If the construction requires construction through or along water courses, investigations may become rather expensive. For river crossings, the decision must be made to either construct in the dry or try underground approaches. With the advent of trenchless technology, pipe drilling and jacking is often more competitive. A decision is required, based on available geologic information, if drilling is performed over water. Barge drilling may be possible in fairly shallow waters where anchorage is relatively easy. However, over deeper open waters, investigations are very expensive. In many cases, investigations can focus on inlet and outlet portals of the jack pit and rely on existing general geology in between. Investigations for bridge piers require a drill hole located where the pier will be constructed.

Again, water flows on rivers and stream courses can be evaluated from gauging station information. Water flow rates and velocities can have a significant impact on constructability and maintenance costs. In some cases, bathometry is required. Side scanning sonar can be useful to detect objects and possible obstructions.

3.2.3 Ground Water

Determination of ground water levels is necessary. This can be done with piezometers or by observing the ground water level in drill holes. If more then one aquifer is present, piezometers should be set in each and isolated to define each aquifer.

In some instances, the soil must be excavated to a level beneath the water table, and the flow of water into the excavation must be eliminated or reduced to an inconsequential amount. To control the inflow of water, a system of drains, pumped wells, and well points must be established either during or, preferably, before removal of the soil. The sides of the excavation are given a slope adequate to maintain stability, or they are braced with some type of support.

Table 7 shows the relationships of permeability, and hydraulic gradient for various soil types. The figure illustrates that the amount of flow can drastically change from dirty to clean soils. These data and other data to follow can be used for estimating dewatering potential.

In general, clean gravels and sands should be dewatered with pump wells, and soils as fine as silts and silty sands can be dewatered with well points or pump wells. If fine, clayey soils are present, often the only alternative is trenching and draining, which take a long time. In cases

Table 7.—Permeability, hydraulic gradient, and groundwater flow-rate relationships for various soil gradations [14]

Soil type	Permeability, cm/s	Gradient i	Time to move 30 cm	n_e
Clean sand	1.0×10^{-2}	0.10 0.01	2.5 hr 25.0 hr	0.30
Silty sand	1.0×10^{-3}	0.10 0.01	1.4 days 14.0 days	0.40
Silt	1.0×10^{-4}	0.10 0.01	14.0 days 140.0 days	0.40
Clayey sand	1.0×10^{-5}	0.10 0.01	174.0 days 4.8 years	0.50
Silty clay	1.0×10^{-6}	0.10 0.01	4.8 years 48.0 years	0.50
Clay (intact)	1.0×10^{-7}	0.10 0.01	48.0 years 480.0 years	0.50

of clay soils, other alternatives, such as shoring, ground stabilization, or mechanical stabilization such as jet grouting, can be used. Dewatering of rock is often done by sumps, and inflow is often controlled by macroscopic features, such as joint patterns.

If a concrete foundation is to be in contact with ground water, a sample of ground water will be required to evaluate the need for sulfate-resistant cement. Sulfate tests should be performed in accordance with the *Concrete Manual* [15]. Pipelines and other metal structures can undergo corrosion. Consult Reclamation's technical guidance on buried metal pipe for corrosion consideration [16]. Foundations should be screened for chlorides, and resistivity measurements made for evaluation of any need for corrosion protection.

3.2.4 Sampling Depth

Figure 6 [1, p. 87] shows the suggested depths of preliminary exploratory holes for various point structures such as small buildings, small pumping plants, transmission towers, and bridge piers. Figure 7 [1, p. 89] shows the suggested depths of preliminary exploratory holes for various line structures such as canals, pipelines, and roads.

3.2.5 Soil Engineering Properties

Many engineering properties of soils can be estimated by knowledge of the soil classification and degree of consistency alone. The exploration team should familiarize themselves with ways to predict necessary soil properties, to make a decision on the amount and degree of investigation.

Soils engineering properties can be obtained from various field and/or laboratory tests. Table 8 lists the type of tests, the parameters obtained from the tests, and the relative costs for the tests.

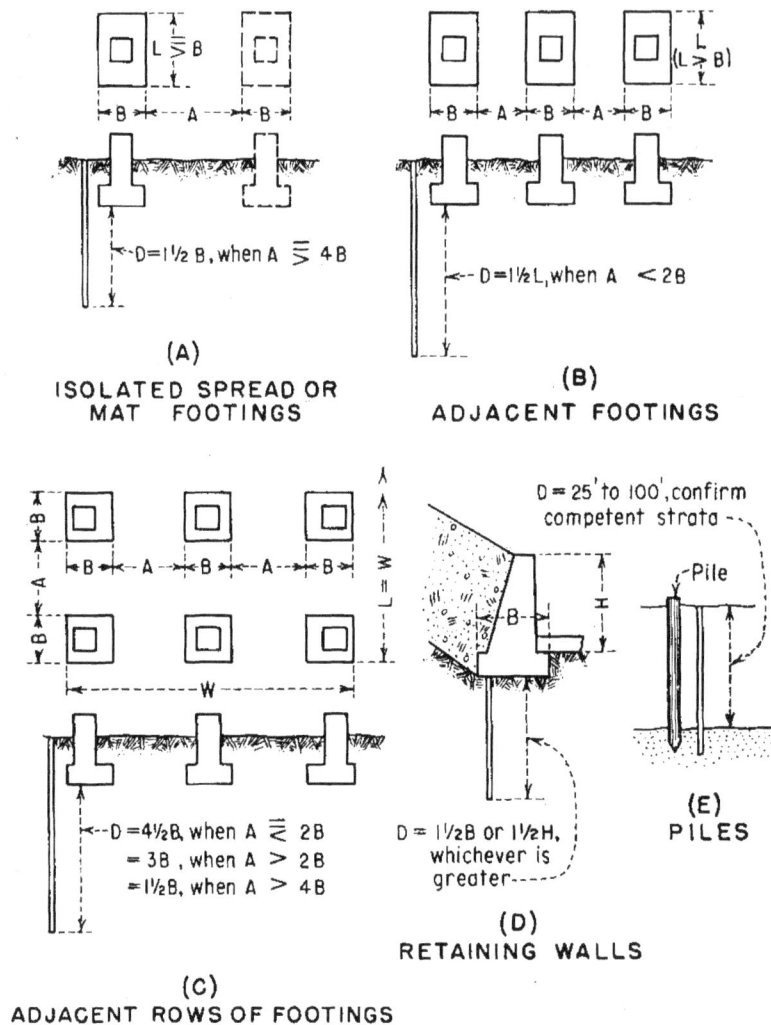

Figure 6—Depth of preliminary exploratory holes for point structures.

Since testing costs change over time, the costs shown in the table are related to the cost of performing and reporting a single visual classification test. For some tests, a cost range is shown. The cost depends on the type of material, the difficulty in processing the material, and additional testing requirements requested. It should be noted that the laboratory tests require some kind of field exploration to obtain the sample. Therefore, there will be an additional cost for obtaining the sample. This is discussed under the *Subsurface Exploration* (3.4) sections.

3.2.5.1 Soil Property Correlations

Generally, technical publications use tables and/or figures to illustrate soil properties correlations to support a hypothesis. A thorough study should be done on how these correlations were obtained, when they used soil parameters from the published tables or figures. For example, the soil parameters for USCS soil type CL shown in table 9 will be discussed.

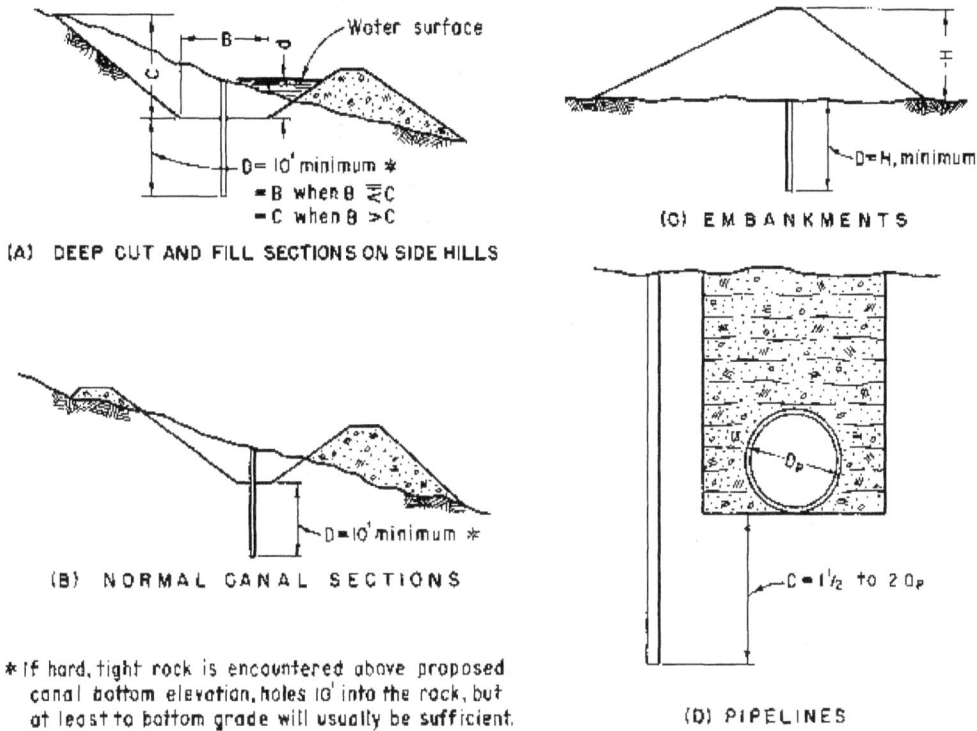

(A) DEEP CUT AND FILL SECTIONS ON SIDE HILLS

(B) NORMAL CANAL SECTIONS

* If hard, tight rock is encountered above proposed canal bottom elevation, holes 10' into the rock, but at least to bottom grade will usually be sufficient.

(C) EMBANKMENTS

(D) PIPELINES

Figure 7.—Depth of preliminary exploratory holes for canal, road, and pipeline alignments.

The soil parameters are for material under the primary group symbol CL, which includes borderline or dual group symbols. Figure 8 shows that the CL group symbol consists of seven group names, depending on the percentage of coarse material (sand and/or gravel). There are values for three tests of plus No. 4 specific gravity. This does not mean that only three of the tests performed for specific gravity had sufficient coarse material for specific gravity testing. It only indicates that for the analysis, it was necessary to have the value for the plus No. 4 specific gravity. The data for the laboratory compaction tests shows a difference of 31.4 lb/ft^3 between the minimum and maximum dry density values and 23 percent between the minimum and maximum optimum moisture contents. The laboratory compaction tests were performed on the minus No. 4 material, which includes sand and fines. The lower test values would indicate finer material and the higher test values would indicate sandy material. The lower optimum moisture content values indicate sandy material, where the higher optimum moisture content values indicate finer material. The amount of coarse material, placement density, and confining pressure affect the shear strength. As the confining pressure increases, the specimen density and friction angle increase. The range of friction angles for the consolidated undrained and consolidated drained tests is between 8 and 34 degrees. The material tested was for use as backfill around shallow structures, which have low confining pressures, and high embankment dams, where the confining pressures are high. A final precaution when using this table is that the table is developed for laboratory compacted soils only. Natural soils could have engineering properties that vary widely from those shown in the table. When using soil parameters from published tables and figures, a thorough

25

Table 8.—Engineering properties tests

Test index	In situ	Laboratory	Test	Parameters obtained	Other tests required	Cost multiplier relative to visual classification
1	X		Visual Classification	USCS Soil Type		1.0
2	X		Inplace Density	Dry Unit Weight, γ_d Moisture Content, W_n		
3	X		Nuclear Moisture-Density	Dry Unit Weight, γ_d Moisture Content, W_n		
4		X	Laboratory Classification Gradation Atterberg Limits	USCS Soil Type Particle Size Distriubition Liquid Limit, LL Plastic Limit, PL Plasticity Index, PI Shrinkage Limit, SL		4 - 6
5		X	Specific Gravity	Specific Gravity, SpG Apparent Bulk SSD Bulk Ovendry Absorption		3 - 4
6		X	Laboratory Compaction Standard	Dry Unit Weight, γ_d Moisture Content, W_o	4, 5	9 - 12
7		X	Laboratory Compaction Large Scale	Dry Unit Weight, γ_d Moisture Content, W_o	4, 5	18 - 20
8		X	Relative Density Coarse Material	Maximum Index Unit Weight, γ_{max} Minimum Index Unit Weight, γ_{min}	4	10 - 12
9		X	Dispersive Analysis Crumb Pinhole Double Hydrometer	Dispersive Grade Dispersive Classification Percent Dispersion	4	2 - 5 9 5 - 6
10	X		Well Permeameter	Average Coefficient of Permeability, k		
11	X		Field Permeability	Coefficient of Permeability, k Flow Rate, q		

Table 8.—Engineering properties tests

Test index	In situ	Laboratory	Test	Parameters obtained	Other tests required	Cost multiplier relative to visual classification
12		X	Permeability Falling Head Constant Head	Coefficient of Permeability, k Settlement	4, 5	15
13		X	Back Pressure Permeability	Coefficient of Permeability, k, versus confining pressure, and hydraulic gradients	4, 5	27 - 38
14		X	Flow Pump Permeability	Coefficient of Permeability, k fine grained soils	4, 5	30 - 37
15		X	Filter	Flow Rate, q suitability of soil or geotextiles as filters	4, 5	25 - 27
16		X	One-Dimensional Consolidat	Coefficient of Consolidation, C_v Settlement	4, 5	12 - 18
17		X	One-Dimensional Expansion	Percent expansion Recompression pressure	4, 5	17
18		X	One-Dimensional Uplift	Uplift Pressure Preconsolidation pressure Percent Expansion	4, 5	20
19	X		Angle of Repose	Cohesionless material lower bound factor of safety	4	12
20		X	Unconfined Compressive Stre	Compressive Strength	4	8 - 17
21	X		Field Vane Shear	Undrained Shear Strength of Soft, Saturated, Fine Grained Soils	4 - 5	
22		X	Laboratory Vane Shear	Undrained Shear Strength of Saturated Cohesive Soils	4 - 5	5
23		X	Direct Shear Fine Grain Material	Drained Shear Strength Angle of Internal Friction, ϕ Cohesion, c	4, 5, (6)	12 - 25

27

Table 8.—Engineering properties tests

Test index	In situ	Laboratory	Test	Parameters obtained	Other tests required	Cost multiplier relative to visual classification
24		X	DirectShear Coarse Grain Material	Peak and Post-Peak Drained Shear Strength Angle of Internal Friction, ϕ Cohesion, c	4, 5(6 or 7)	15 - 28
25		X	Repeated Direct Shear	Residual Shear Strength Angle of Internal Friction, ϕ Cohesion, c	4, 5	7 - 12
26		X	Rotational Shear	Residual Shear strength Angle of Internal Friction, ϕ Cohesion, c	4, 5	30
27		X	Simple Shear	Low strain shear strength	4, 5	33.3
28		X	Initial Capillary Pressure	Initial Capillary Pressure (suction) of partially saturated soils	4, 5	16.7
29		X	Coefficient of Earth Pressureat Rest, K_O	Coefficient of Earth Pressure at Rest, K_O Poisson's Ratio, μ Modulus of Deformation, E_d	4, 5, (6 or 7)	30 - 37
30		X	Unconsolidated-UndrainedTriaxial Shear, UU	Shear Strength Total or Effective Stress Analysis Angle of Internal Friction, ϕ Cohesion, c	4, 5, (6 or 7)	12 - 48
31		X	Consolidated-UndrainedTriaxial Shear, CU	Shear Strength Total or Effective Stress Analysis Angle of Internal Friction, ϕ Cohesion, c	4, 5, (6 or 7)	18 - 93
32		X	Consolidated-Drained Triaxial Shear, CD	Shear Strength Total Stress Analysis Angle of Internal Friction, ϕ Cohesion, c	4, 5, (6 or 7)	18 - 93

Table 9.—Average engineering properties of compacted soils from the western United States. Last updated October 6, 1982.

USGS soil type	Spec. gravity		Compaction				Shear strength				Values listed
			Laboratory		Index unit weight		Avg. placement		Effective stress		
	No. 4 minus	No. 4 plus	Max. unit weight, lb/ft^2	Optimum moisture content, %	Max., lb/ft^3	Min., lb/ft^3	Unit weight, lb/ft^3	Moisture content, %	c', lb/in^2	ϕ', degrees	
GW	2.69	2.58	124.2	11.4	133.6	108.8	-	-			Average of all values
	0.02	0.08	3.2	1.2	10.4	10.2	-	-			Standard deviation
	2.65	2.39	119.1	9.9	113.0	88.5	-				Minimum value
	2.75	2.67	127.5	13.3	145.6	132.9	-	-		-	Maximum value
	16	9	5		16		0				Total number of tests
GP	2.68	2.57	121.7	11.2	137.2	112.5	127.5	6.5	5.9	41.4	Average of all values
	0.03	0.07	5.9	2.2	6.3	8.3	7.2	1.2		2.5	Standard deviation
	2.61	2.42	104.9	9.1	118.3	85.9	117.4	5.3	5.9	38.0	Minimum value
	2.76	2.65	127.7	17.7	148.8	123.7	133.9	8.0	5.9	43.7	Maximum value
	35	12	15		34		3				Total number of tests
GM	2.73	2.43	113.3	15.8	132.0	108.0	125.9	10.3	13.4	34.0	Average of all values
	0.07	0.18	11.5	5.8	3.1	0.2	0.9	1.2	3.7	2.6	Standard deviation
	2.65	2.19	87.0	5.8	128.9	107.8	125.0	9.1	9.7	31.4	Minimum value
	2.92	2.92	133.0	29.5	135.1	108.1	126.9	11.5	17.0	36.5	Maximum value
	34	17	36		2		2				Total number of tests
GC	2.73	2.57	116.6	13.9			111.1	15.9	10.2	27.5	Average of all values
	0.08	0.21	7.8	3.8			10.4	1.6	1.5	7.2	Standard deviation
	2.67	2.38	96.0	6.0	-		96.8	11.2	5.0	17.7	Minimum value
	3.11	2.94	129.0	23.6	-		120.9	22.2	16.0	35.0	Maximum value
	34	6	37		0		3				Total number of tests
SW	2.67	2.57	126.1	9.1	125.0	99.5					Average of all values
	0.03	0.03	6.0	1.7	6.0	7.1					Standard deviation
	2.61	2.51	118.1	7.4	116.7	87.4	-	-			Minimum value
	2.72	2.59	135.0	11.2	137.8	109.8	-				Maximum value
	13	2	1		12		0				Total number of tests
SP	2.65	2.62	115.6	10.8	115.1	93.4	103.4	5.4	5.5	37.4	Average of all values
	0.03	0.10	9.7	2.0	7.2	8.8	14.6	-	3.0	2.0	Standard deviation
	2.60	2.52	106.5	7.8	105.9	78.2	88.8	5.4	2.5	35.4	Minimum value
	2.77	2.75	134.8	13.4	137.3	122.4	118.1	5.4	8.4	39.4	Maximum value
	36	3	7		39		2				Total number of tests
SM	2.68	2.18	116.6	12.5	110.1	84.9	112.0	12.7	6.6	33.6	Average of all values
	0.06	0.11	8.9	3.4	8.7	7.9	11.1	5.4	5.6	5.7	Standard deviation
	2.51	2.24	92.9	6.8	88.5	61.6	91.1	1.6	0.2	23.3	Minimum value
	3.11	2.63	132.6	25.5	122.9	97.1	132.5	25.0	21.2	45.0	Maximum value
	149	9	123		21		17				Total number of tests
SC	2.69	2.17	118.9	12.4	-		115.6	14.2	5.0	33.9	Average of all values
	0.04	0.18	5.9	2.3	14.1			5.7	2.5	2.9	Standard deviation
	2.56	2.17	104.3	6.7	-		91.1	7.5	0.7	28.4	Minimum value
	2.81	2.59	131.7	18.2	-		131.8	22.7	8.5	38.3	Maximum value
	88	4	73		0		10				Total number of tests

Table 9.—Average engineering properties of compacted soils from the western United States. Last updated October 6, 1982.

USGS soil type	Spec. gravity		Compaction				Shear strength				Values listed
			Laboratory		Index unit weight		Avg. placement		Effective stress		
	No. 4 minus	No. 4 plus	Max. unit weight, lb/ft²	Optimum moisture content, %	Max., lb/ft³	Min., lb/ft³	Unit weight, lb/ft³	Moisture content, %	c', lb/in²	ϕ', degrees	
ML	2.69	-	103.3	19.7	-		98.9	22.1	3.6	34.0	Average of all values
	0.09		10.4	5.7	-		11.5	8.9	4.3	3.1	Standard deviation
	2.52	-	81.6	10.6			80.7	11.1	0.1	25.2	Minimum value
	3.10	-	126.0	34.6	-		119.3	40.3	11.9	37.7	Maximum value
	65	0	39		0		14				Total number of tests
CL	2.71	2.59	109.3	16.7	-		106.5	17.7	10.3	25.1	Average of all values
	0.05	0.13	5.5	2.9			7.8	5.1	7.6	7.0	Standard deviation
	2.56	2.42	90.0	6.4	-		85.6	11.6	0.9	8.0	Minimum value
	2.87	2.75	121.4	29.2	-		118.7	35.0	23.8	33.8	Maximum value
	270	3	221		0		31				Total number of tests
MH	2.79	-	85.1	33.6	-	-					Average of all values
	0.25		2.3	1.6	-	-	-				Standard deviation
	2.47		82.9	31.5	-						Minimum value
	3.50		89.0	35.5	-						Maximum value
	10	0	5		0		0				Total number of tests
CH	2.73	-	95.3	25.0			93.6	25.7	11.5	16.8	Average of all values
	0.06		6.6	5.4			8.1	5.7	7.4	7.2	Standard deviation
	2.51	-	82.3	16.6	-	-	79.3	17.9	1.5	4.0	Minimum value
	2.89	-	107.3	41.8			104.9	35.3	21.5	27.5	Maximum value
	74	0	36		0		12				Total number of tests

understanding of the material being used and how it will be used is necessary.

Table 10 [17] shows typical properties of compacted soils for classification group symbols. The number of tests used for this table is not reported.

3.2.5.2 Permeability

Permeability is important in hydraulic structures like Reclamation's, because many are built to be water barriers of some kind. In granular soils it is surprising, however, that it takes only small amounts of fines (silt or clay) to reduce permeability.

Table 11 and figures 9 through 12 show permeability and drainage ranges for various soil types. It has to be assumed that the results are from laboratory testing. The hydraulic gradient is not indicated. A use of the figures would be to compare the parameters of all these figures and any others that can be found for a soil type of interest. That would narrow the permeability value range for design.

Soils such as clean gravels, sands, and some uncompacted silts have high permeability, although these soils, such as gravelly soils, may be very desirable regarding stability and low

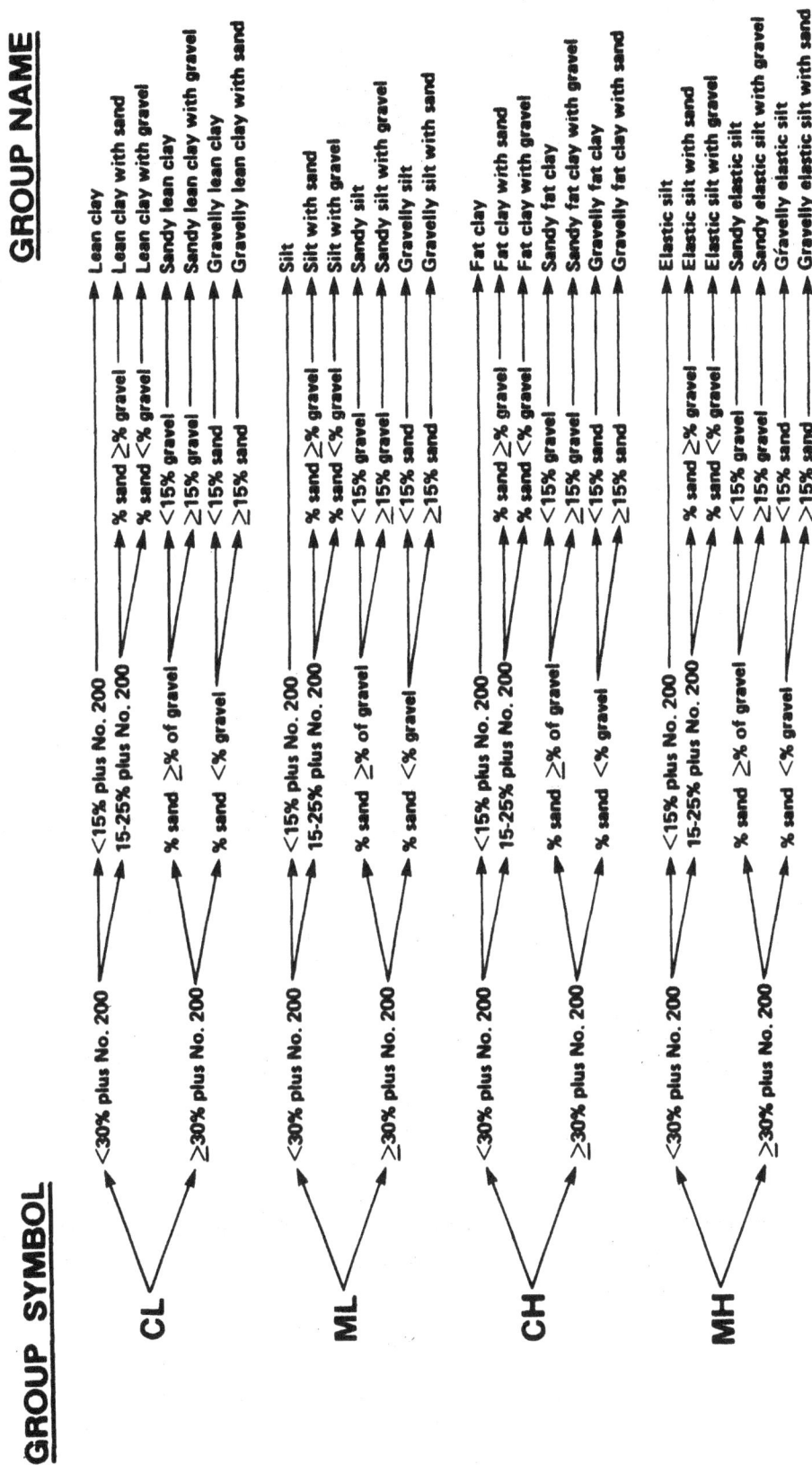

Figure 8.—Flowchart for identifying organic fine grained soil (50% or more fines)—visual-manual method.

Table 10.—Typical properties of compacted soils [17]

Group Symbol	Soil Type	Range of Max. Dry Unit Weight, lb/ft³	Range of Optimum Moisture, %	Typical Value of Compression — Percent of Original Height — At 1.4 t/ft² (20 lb/in²)	At 3.6 t/ft² (50 lb/in²)	Typical Strength Characteristics — Cohesion (as compacted) lb/ft²	Cohesion (saturated) lb/ft²	φ (Effective Stress Envelope Degrees)	Tan φ	Typical Coefficient of Permeability ft/min	Range of CBR Values	Range of Subgrade Modulus k lb/in³
GH	Well graded clean gravels, gravel-sand mixtures	125 - 135	11 - 8	0.3	0.6	0	0	>38	>0.19	5×10^{-2}	40 - 80	300 - 500
GP	Poorly graded clean gravels, gravel-sand mix	115 - 125	14 - 11	0.4	0.9	0	0	>31	>0.14	10^{-1}	30 - 60	250 - 400
GH	Silty gravels, poorly graded gravel-sand-silt	120 - 135	12 - 8	0.5	1.1	. . . . .	. . . . .	>34	>0.67	$>10^{-6}$	20 - 60	100 - 400
GC	Clayey gravels, poorly graded gravel-sand-clay	115 - 130	14 - 9	0.7	1.6	. . . . .	. . . . .	>31	>0.60	$>10^{-1}$	20 - 40	100 - 300
SW	Well graded clean sands, gravelly sands	110 - 130	16 - 9	0.6	1.2	0	0	18	0.79	$>10^{-3}$	20 - 40	200 - 300
SP	Poorly graded clean sands, sand-gravel mix	100 - 120	21 - 12	0.8	1.4	0	0	11	0.74	$>10^{-3}$	10 - 40	200 - 300
SH	Silty sands, poorly graded sand-silt mix	110 - 125	16 - 11	0.8	1.6	1050	420	34	0.67	$5 \times >10^{-5}$	10 - 40	100 - 300
SH-SC	Sand-silt clay mix with slightly plastic fines	110 - 130	15 - 11	0.8	1.4	1050	300	13	0.66	$2 \times >10^{-6}$	5 - 30	100 - 300
SC	Clayey sands, poorly graded sand-clay-mix	105 - 125	19 - 11	1.1	2.2	550	230	31	0.60	$5 \times >10^{-7}$	5 - 20	100 - 300
HL	Inorganic silts and clayey silts	95 - 120	24 - 12	0.9	1.7	400	190	32	0.62	$>10^{-5}$	15 or less	100 - 200
HL-CL	Mixture of Inorganic silt and clay	100 - 120	22 - 12	1.0	2.2	350	460	32	0.62	$5 \times >10^{-1}$		
CL	Inorganic clays of low to medium plasticity	95 - 120	24 - 12	1.3	2.5	800	210	28	0.54	$>10^{-1}$	15 or less	50 - 200

Table 10.—Typical properties of compacted soils [17]

Group Symbol	Soil Type	Range of Max. Dry Unit Weight, lb/ft³	Range of Optimum Moisture, %	Typical Value of Compression At 1.4 t/ft² (20 lb/in²) Percent of Original Height	Typical Value of Compression At 3.6 t/ft² (50 lb/in²) Percent of Original Height	Cohe-sion (as com-pacted) lb/ft²	Cohe-sion (satu-rated) lb/ft²	φ (Effective Stress Envelope Degrees)	Tan φ	Typical Coefficient of Permeability ft/min	Range of CBR Values	Range of Subgrade Modulus k lb/in³
OL	Organic silts and silt- clays, low plasticity	80 - 100	33 - 21	. . . . .	. . . . .	. . . . .	. . . . .	. . . . .	. . . . .	. . . . .	5 or less	100
MI	Inorganic clayey silts elastic silts	70 - 95	40 - 24	2.0	3.8	500	420	25	0.47	$5 \times {>}10^{-1}$	10 or less	50 - 100
CH	Inorganic clay, of high plasticity clays	75 - 105	36 - 19	2.6	3.9	2150	230	19	0,35	${>}10^{-7}$	15 or less	50 - 150
OH	Organic clays and silty	65 - 100	45 - 21	. . . . .	. . . . .	. . . . .	. . . . .	. . . . .	. . . . .	. . . . .	5 or less	25 - 100

Notes:
1. All properties are for condition of "Standard Proctor" maximum density, except values of k and C8R which ere for "modified Proctor" maximum density.
2. Typical stength characteristics are for effective strength envelopes and are obtained from Reclamation data.
3. Compression values are for vertical loading with complete lateral confinement.
4. (>) indicates that typical property is greater then the value shown. (. . . .) indicates insufficient data available for an estimate.

33

Table 11.—Typical permeability coefficients for various materials [14]

| | Particle-size range | | | | "Effective" size | | Permeability coefficient k | | |
| | Inches | | Millimeters | | | | | | |
	D_{max}	D_{min}	D_{max}	D_{min}	D_{20}, in	D_{18}, mm	ft/year	ft/month	cm/s
Turbulent flow									
Derrick stone	120	36			48		100×10^6	100×10^5	100
One-man stone	12	4			6		30×10^6	30×10^5	30
Clean, fine to coarse gravel	3	¼	80	10	½		10×10^6	10×10^5	10
Fine, uniform gravel	⅜	1/16	8	1.5	⅛		5×10^6	5×10^5	5
Very coarse, clean, uniform sand	⅛	1/32	3	0.8	1/16		3×10^6	3×10^5	3
Laminar flow									
Uniform, coarse sand	⅛	1/64	2	0.5		0.6	0.4×10^6	0.4×10^5	0.4
Uniform, medium sand			0.5	0.25		0.3	0.1×10^6	0.1×10^5	0.1
Clean, well-graded sand and gravel			10	0.05		0.1	0.01×10^6	0.01×10^5	0.01
Uniform, fine sand			0.25	0.05		0.06	4000	400	40×10^{-4}
Well-graded, silty sand and gravel			5	0.01		0.02	400	40	4×10^{-4}
Silty sand			2	0.005		0.01	100	10	10^{-4}
Uniform silt			0.05	0.005		0.006	50	5	0.5×10^{-4}
Sandy clay			1.0	0.001		0.002	5	0.5	0.05×10^{-4}
Silty clay			0.05	0.001		0.0015	1	0.1	0.01×10^{-4}
Clay (30 to 50% clay sizes)			0.05	0.0005		0.0008	0.1	0.01	0.001×10^{-4}
Colloidal clay ($-2\mu \leq 50\%$)			0.01	10 Å		40 Å	0.001	10^{-4}	10^{-9}

Coefficient of permeability, k, in cm/s (log scale)

10^2	10^1	1.0	10^{-1}	10^{-2}	10^{-3}	10^{-4}	10^{-5}	10^{-6}	10^{-7}	10^{-8}	10^{-9}

Drainage

Good	Poor	Practically Impervious

Soil types

- Clean gravel
- Clean sands, clean sand and gravel mixtures
- Very fine sands, organic and inorganic silts, mixtures of sand silt and clay, glacial till, stratified clay deposits, etc.
- "Impervious" soils modified by effects of vegetation and weathering
- "Impervious" soils, e.g., homogeneous clays below zone of weathering

Direct determination of k

- Direct testing of soil in its original position—pumping tests. Reliable if properly conducted. Considerable experience required
- Constant-head permeameter. Little, experience required

Indirect determination of k

- Falling-head permeameter. Reliable. Little experience required
- Falling-head permeameter. Unreliable. Much experience required
- Falling-head permeameter. Fairly reliable. Considerable experience necessary
- Computation from grain-size distribution. Applicable only to clean cohesionless sands and gravels
- Computation based on results of consolidation tests. Reliable. Considerable experience required

Figure 9.—Coefficient of permeability, k, in cm/s (log scale) [18].

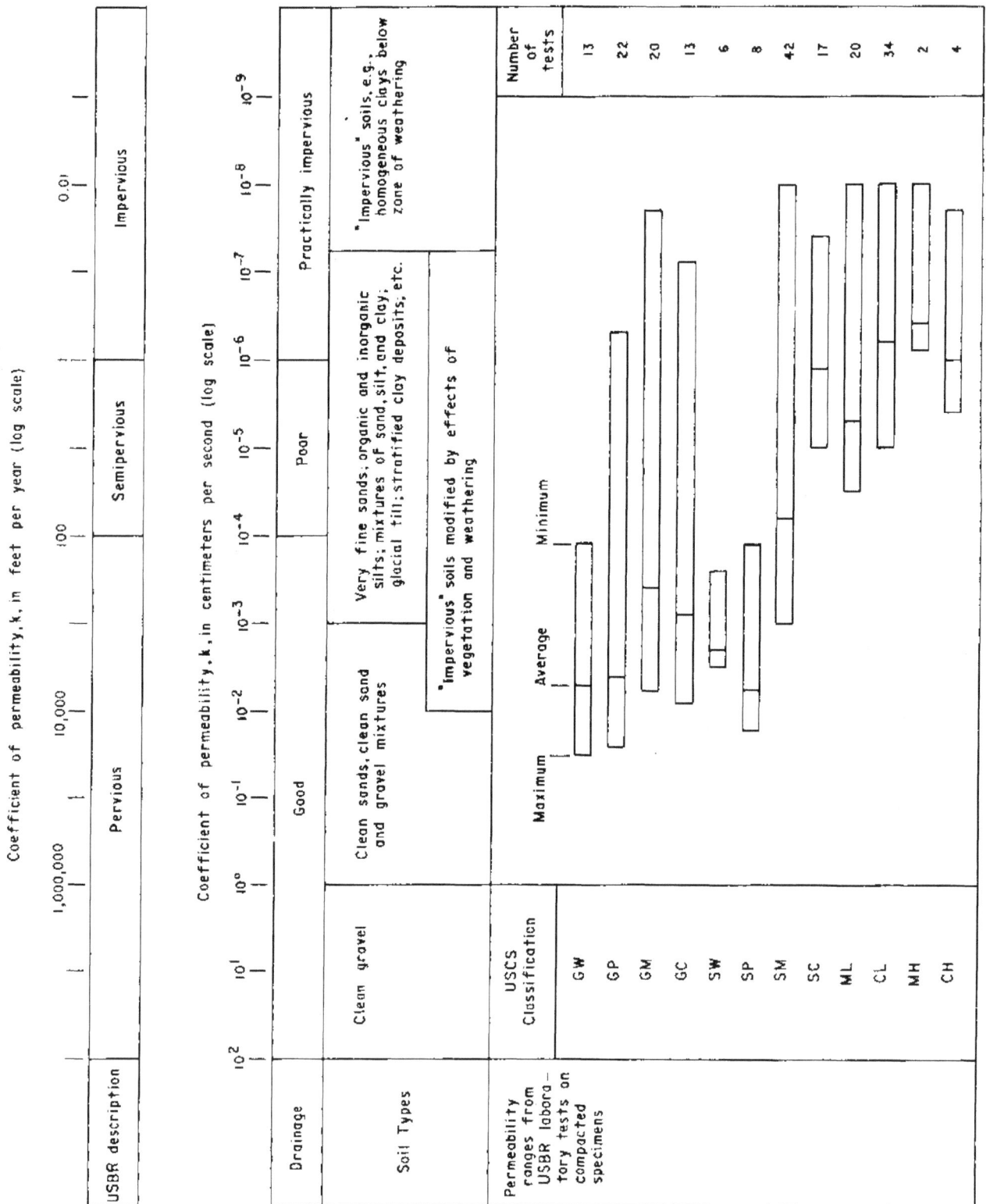

Figure 10.—Permeability ranges for soils [18].

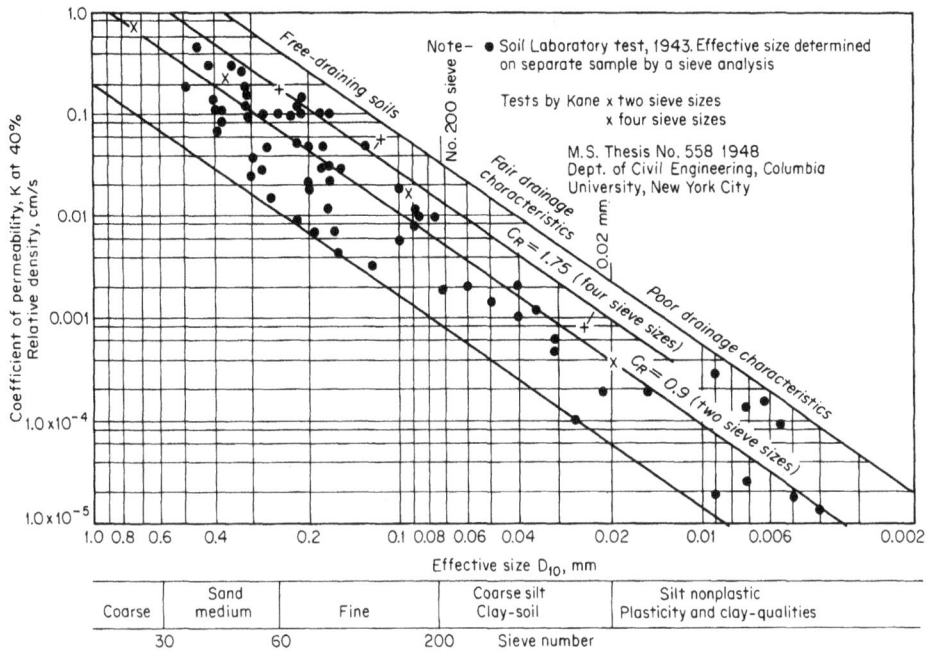

Figure 11.—Relationships between permeability and Hazen's effective size D_{10}. Coefficient of permeability reduced to basis of 40% D_R by fig. 12 [14].

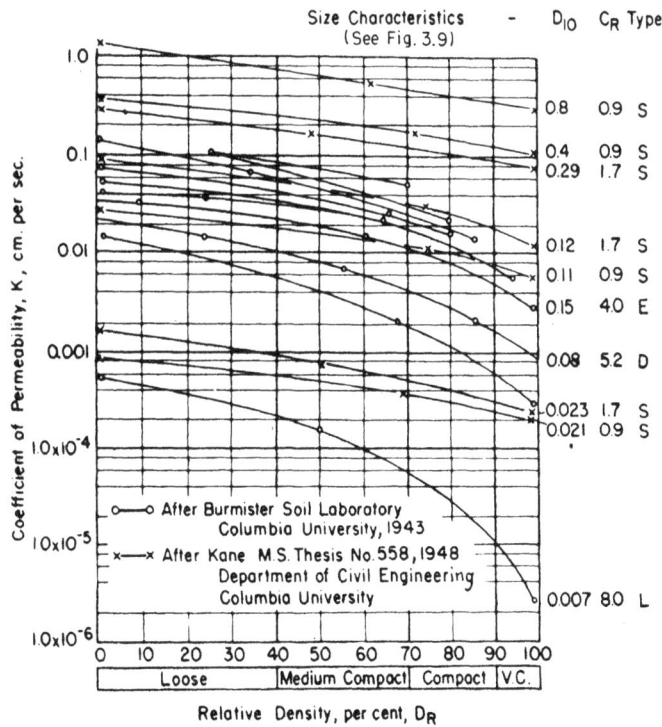

Figure 12.—Permeability-relative density relationships [14].

settlement. When permeability becomes excessive, water movement can remove particles, and this piping action will undermine structures. In canals, piping may cause sinkholes in addition to the objectionable loss of water. Blocking the movement of fine soil with layers of selected soils in a filter corrects the problem of high permeability and piping. However, from a soil mechanics standpoint, some clay content is the most desirable natural means of keeping permeability under control.

Figure 13 [19] clearly shows how soils have permeabilities ranging over 10 orders of magnitude. Using soil classification and observation of the soil structure, one ought to be able to estimate the permeability within one or two orders of magnitude and this should be sufficient for most investigations. Clean, granular soils of high and medium permeability will not vary much more than one order of magnitude. However, for the clayey soils in very low to impermeable zones, permeability can range up to several orders of magnitude, depending on void ratio or "structure" in the soil. Secondary structural features, such as fissuring and desiccation cracking in clay, often control the permeability.

It is very difficult to impossible to obtain undisturbed samples of clean sands and gravels. Fortunately, the permeability of these soils can be estimated readily by charts, or by equations such as Hazen's formula. If the estimated values are critical, falling head permeability tests can be performed on remolded disturbed sand samples.

For excavations below the ground water level, dewatering is likely to be required during construction, and the investigation should give the contractor information on the soil types and water levels to be encountered. Borings performed onsite should be allowed to stand

38

open for accurate water level information. For smaller structures, expensive dewatering systems may not be required. Often excavations can be enclosed in sheet piling or a slurry trench, or pumping can drain the excavation. Most Reclamation specifications require that excavation be performed "in the dry." However, if the appropriate soil information is not available, difficulties may be encountered. For example, soft clays are difficult to dewater, and excavation stability could be worsened. Aquifer tests provide the best data for dewatering, yet they are very expensive and are mostly performed for major structures.

3.2.5.3 Compressibility

Some settlement can always be expected for foundations resting on soil. However, it is important to keep these settlements within tolerable limits. When "no settlement" is permitted, the foundation needs to be placed on solid rock or on piers, piles, or caissons. Some settlement is not objectionable if it is not of appreciable differential amounts beneath the structure. The Leaning Tower of Pisa is a classic example of differential settlement. Typical geotechnical design for foundations is governed by required settlements of less than 1 inch, and often the settlement criteria govern allowable pressures compared to bearing capacity requirements. For most small structures, contact pressures are not large (generally less than 1-2 t/ft^2). If settlement is a concern, the structure can be placed below ground, such that the net pressure change from excavation to structure placement is zero. If conditions are variable under the structure, the foundation can be overexcavated and replaced with a layer of uniform, compacted fill to result in more uniform settlement.

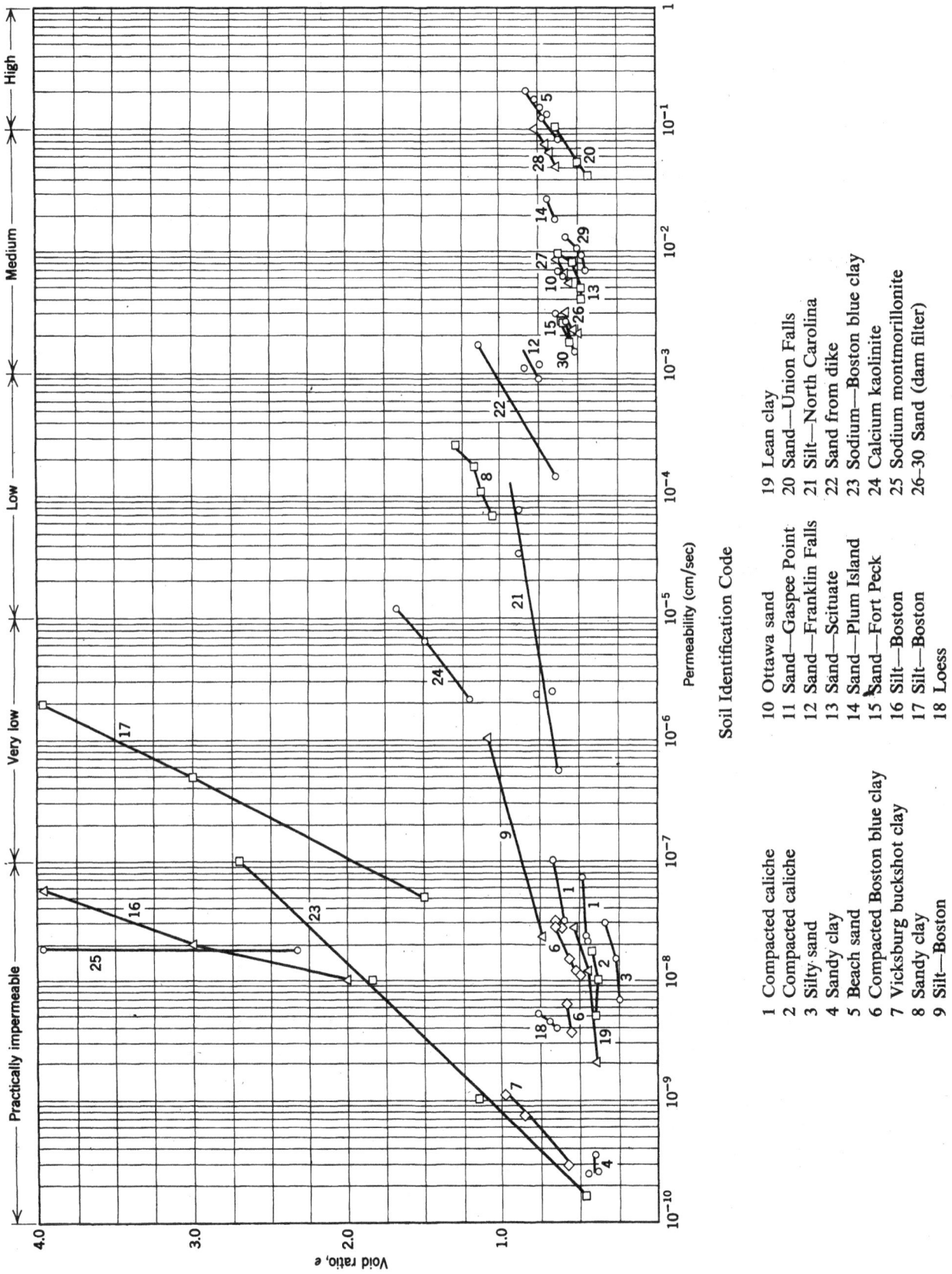

Soil Identification Code

1 Compacted caliche	10 Ottawa sand	19 Lean clay
2 Compacted caliche	11 Sand—Gaspee Point	20 Sand—Union Falls
3 Silty sand	12 Sand—Franklin Falls	21 Silt—North Carolina
4 Sandy clay	13 Sand—Scituate	22 Sand from dike
5 Beach sand	14 Sand—Plum Island	23 Sodium—Boston blue clay
6 Compacted Boston blue clay	15 Sand—Fort Peck	24 Calcium kaolinite
7 Vicksburg buckshot clay	16 Silt—Boston	25 Sodium montmorillonite
8 Sandy clay	17 Silt—Boston	26-30 Sand (dam filter)
9 Silt—Boston	18 Loess	

Figure 13.—Permeability test data [20].

39

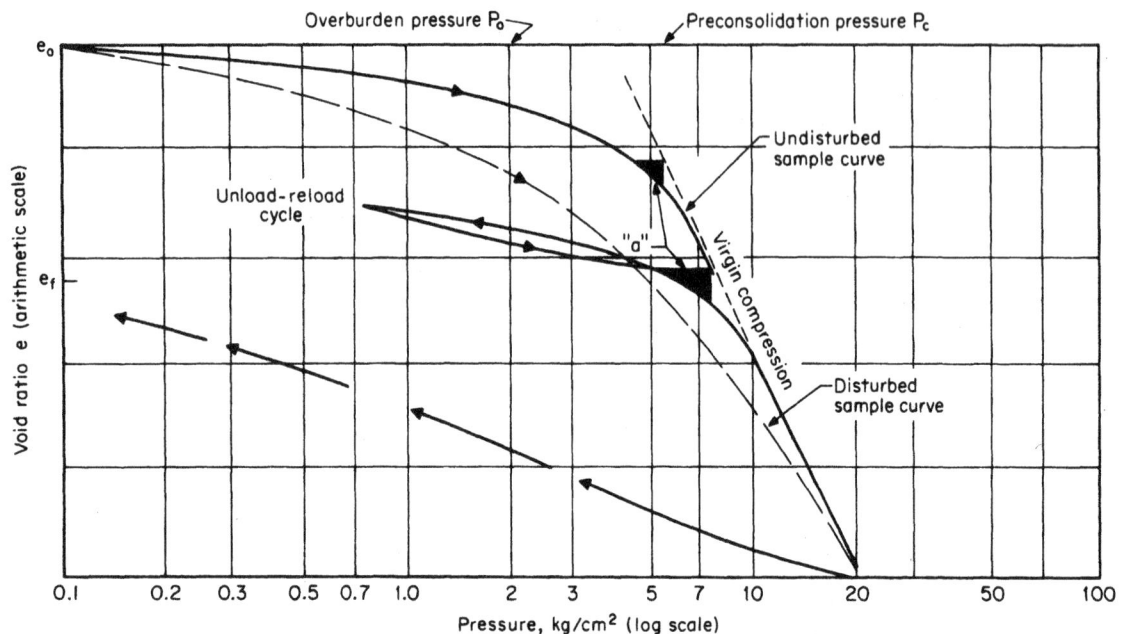

Figure 14.—Consolidation test: pressure-void ratio curve (e-log-p) [14].

In hydraulic structures, it may be better to have the structure settle with the foundation to prevent underseepage than to have it supported on piles or piers. The penetration resistance test would be useful to evaluate the denseness and firmness of the foundations when more detailed laboratory consolidation tests are not warranted. The principal ways to control settlement are (1) increase footing size, (2) compact earth pads, (3) do the construction in stages, (4) use ground improvement methods (dynamic compaction, jet grouting, etc.), or (5) use piles.

Soil, as found in nature, has a certain amount of strength that can be destroyed by disturbance. This is very true of highly compressed clays. Figure 14 is an example of a load consolidation test in which an undisturbed sample of soil shows considerable resistance to settlement to loads up to 5 kg/cm^2, but under higher loads this natural strength breaks down. When the same soil is disturbed and recompacted, it does

not have as much resistance to settlement, as shown by the dashed line.

In the arid western states, structures are often founded above the water table on dry, desiccated soils. When performing investigations of these soils, it should be kept in mind that the strength of the soils will change if wetted. For example, the initial load settlement relationship shown on figure 14 might be typical of a clay soil overconsolidated by desiccation. Upon wetting, the soil will likely be more compressible and possibly expansive or collapsible.

Figures 15 through 17 show pressure void ratio curves for various soils.

Fairly clean, predominantly quartz and feldspar sand and gravel soils have very low compressibility. Settlement under lightly loaded structures would be minimal. However it is important to note if any compressible grains (for example glauconitic or carbonate particles)

40

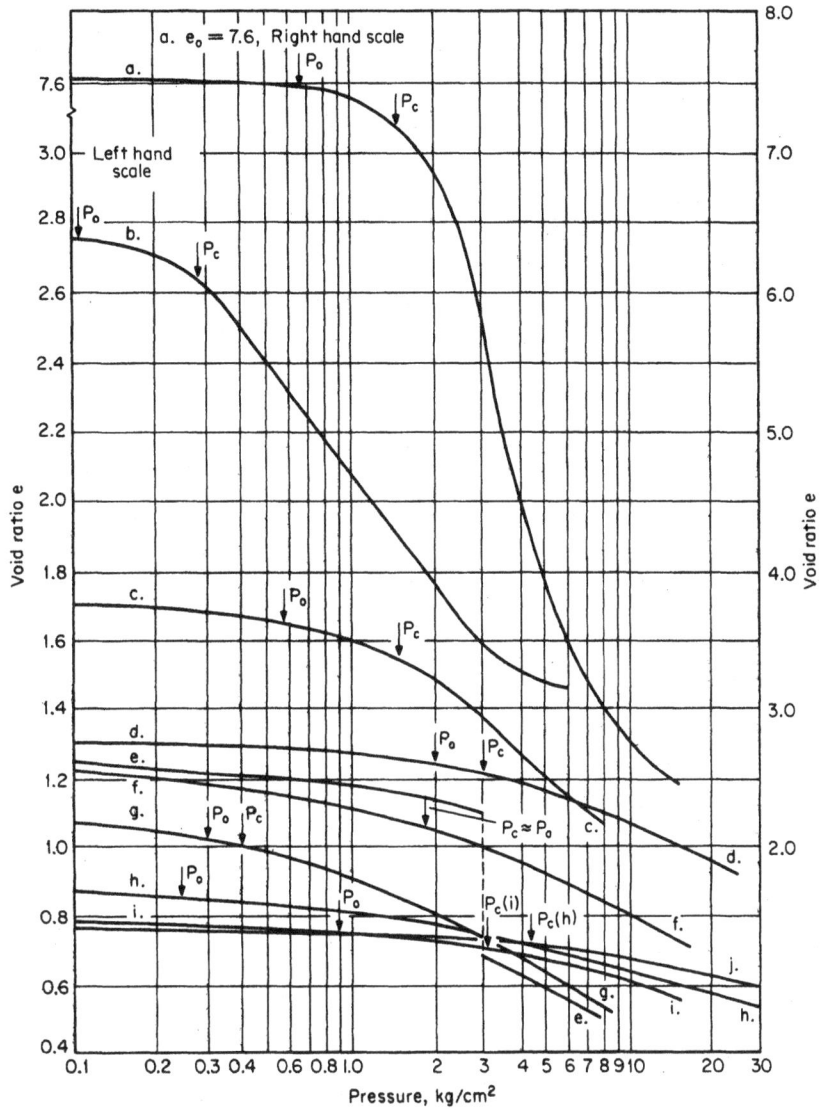

Material	γ_d	w	LL	PI	Material	γ_d	w	LL	PI
a. Soft silty clay (lacustrine, Mexico)	0.29	300	410	260	f. Medium sandy, silty clay (residuual gneiss, Brazil)	1.29	38	40	16
b. Soft organic silty clay (alluvium, Brazil)	0.70	92			g. Soft silty clay (alluvium, Texas)		32	48	33
c. Soft organic silty clay (backswamp, Georgia)	0.96	65	76	31	h. Clayey silt, some fine sand (shallow alluvium, Georgia)	1.46	29	53	24
d. Stiff clay varve (glaciolacustrine, New York)		46	62	34	i. Stiff clay (Beaumornt clay, Texas)	1.39	29	81	55
e. "Porous" clay (residual, Brazil)	1.05	32	43	16	j. Silt varve (glaciolacustrine, New York)				

Figure 15.—Typical pressure-void ratio curves for various clay soils [14].

41

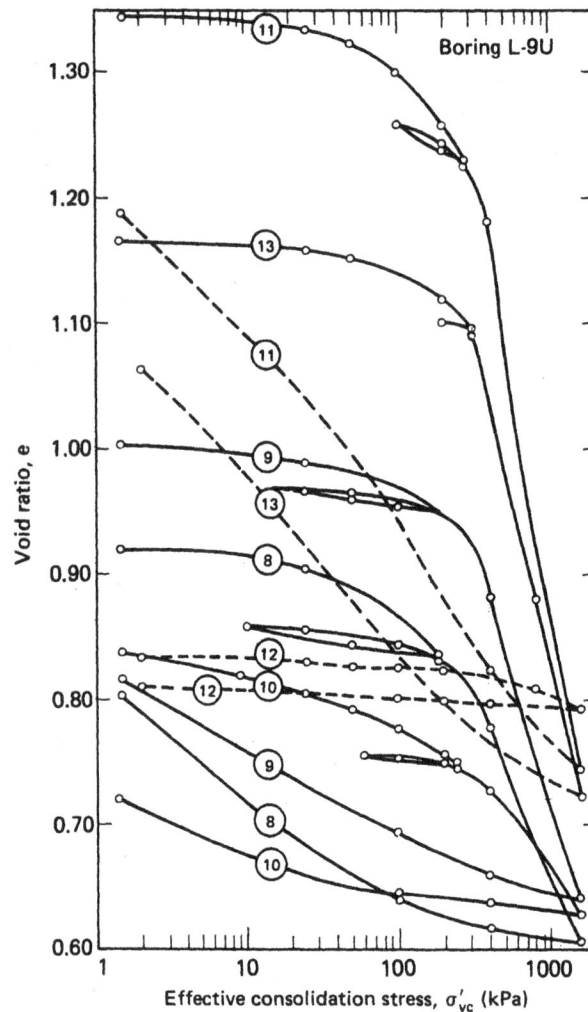

Test No.	Elev. (m)	Classification	Atterberg limits			w_n (%)	e_o	σ'_{vo} (kPa)	σ'_p (kPa)	C_c
			LL	PL	PI					
8	-8.6	CL-clay, soft	41	24.	17	34.0	0.94	160	200	0.34
9	-9.8	CL-clay, firm	50	23	27	36.4	1.00	170	250	0.44
10	-17.1	ML-sandy silt	31	25	6	29.8	0.83	230	350	0.16
11	-20.1	CH-clay, soft	81	25	56	50.6	1.35	280	350	0.84
12	-23.2	SP-sand	Nonplastic			27.8	0.83	320		
13	-26.2	CH-clay w/silt strata	71	28	43	43.3	1.17	340	290	0.52

Figure 16.–Nearly normally consolidated clays and silts [13].

or inclusions are present. One common inclusion element in sands is mica, and its presence can greatly increase the compressibility as shown on figure 17.

Tables 12 and 13 tabulate some compression index, C_c, values for various clayey and silty soils. C_c is the slope of the virgin compression curve as shown on figure 17. C_c is used to

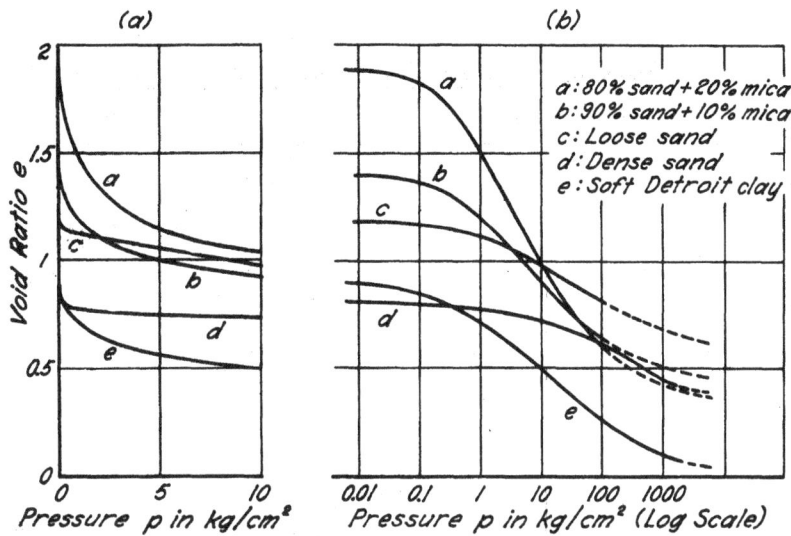

Figure 17.—(a) Typical e-p curves. (b) Corresponding e-log p curves representing results of compression tests on laterally confined laboratory soil aggregates [18].

Table 12.—Typical values of the compression index C_c [13]

Soil	C_c
Normally consolidated medium sensitive clays	0.2 to 0.5
Chicago silty clay (CL)	0.15 to 0.3
Boston blue clay (CL)	0.3 to 0.5
Vicksburg buckshot clay (CH)	0.5 to 0.6
Swedish medium sensitive clays (CL-CH)	1 to 3
Canadian Leda clays (CL-CH)	1 to 4
Mexico City clay (MH)	7 to 10
Organic clays (OH)	44 and up
Peats (Pt)	10 to 15
Organic silt and clayey silts (ML-MH)	1.5 to 4.0
San Francisco Bay mud (CL)	0.4 to 1.2
San Francisco Old Bay clays (CH)	0.7 to 0.9
Bangkok clay (CH)	0.4

determine settlement characteristics of the material. Note that for clays, knowledge of the in situ water content and/or Atterberg limits alone can be used to estimate C_c.

The void ratio, e_0, shown in table 13, of the in-place material can be easily obtained from an in-place density test and measurement of specific gravity.

From the consolidation tests, time-consolidation curves are plotted for normal pressures as shown on figure 18. From these curves, the coefficients of consolidation, C_v, are computed. With these values, the time required for settlement for various loads can be computed. Time consolidation data are important data to use when fill loading is used to consolidate soft clay deposits.

43

Table 13.—Some empirical equations for C_c and $C_{c\epsilon}$ [13]

Equation	Regions of Applicability
$C_c = 0.007 \cdot (LL - 7)$	Remolded clays
$C_{c\epsilon} = 0.208e_o + 0.0083$	Chicago clays
$C_c = 17.66 \times 10^{-5}w_n^2 + 5.93 \times 10^{-3}w_n - 1.35 \times 10$	Chicago clays
$C_c = 1.15(e_o - 0.35)$	All clays
$C_c = 0.30(e_o - 0.27)$	Inorganic, cohesive soil; silt; some clay; silty clay; clay
$C_c = 1.15 \times 10^{-2}w_n$	Organic soils—meadow mats, peats, and organic silt and clay
$C_c = 0.75(e_o - 0.50)$	Soils of very low plasticity
$C_{c\epsilon} = 0.156e_o + 0.0107$	All clays
$C_c = 0.01w_n$	Chicago clays

Note: w_n = natural water content

For fairly pervious sands and gravels, settlement occurs rapidly and can be assumed to occur during construction. Settlement in sands and gravels of features such as earth fill embankments are not a concern, because the embankment can be built to final line and grade. However, settlement of sands during construction for a pumping plant may be a concern, because the line and grade of the piping must be maintained at design levels. Settlements of sands are evaluated by determining the stiffness of the sand through penetration resistance tests.

Loose or collapsible soils settle excessively when wetted. The best remedial measure is to collapse them by wetting, excavate and replace, or compact them before building a structure on them.

3.2.5.4 Strength

The structure foundation must be safe against punching into the ground. This may involve both shearing properties and consolidation properties. Saturated, weak clay can be visualized as resulting in settlement of the soil under the foundation, accompanied by bulging of the soil out from around the foundation. A large, rigid structure sometimes causes catastrophic failures because of this shearing weakness; however, when a soil is found to be firm, it can resist this weakness. Penetration resistance, vane tests, density tests, and laboratory testing of undisturbed samples are ways of evaluating this firmness. Improving the foundation with regard to bearing capacity problems can be accomplished by (1) deeper foundations, (2) overexcavation and refill, (3) piles to firm material, (4) ground

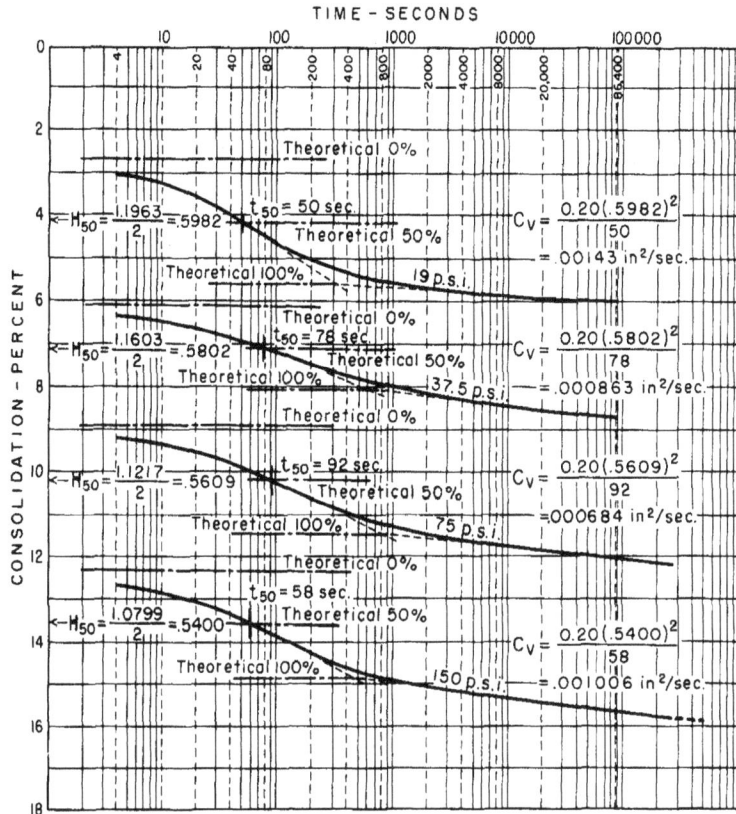

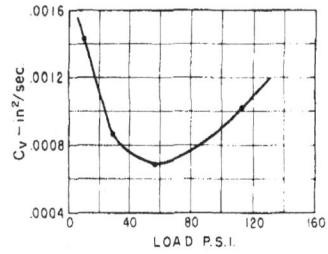

Figure 18—Determination of the coefficient of consolidation, C_v, for the typical example.

improvement, and (4) larger foundation areas to reduce load. Table 14 summarizes methods for strengthing foundations.

The shearing strength of soil depends on many factors. The primary consideration is the loading rate and drainage of excess water pressures that can occur as the soil is loaded. The basic strength equation according to Mohr-Colomb theory is;

$$S = C + \sigma\!\!\!\!/ \tan \phi$$

where

S = unit shearing resistance on the shear surface

C = cohesion, strength component independent of pore pressure

$\sigma\!\!\!\!/$ = effective stress on the shear surface (total stress minus pore water pressure)

ϕ = angle of internal friction of the soil

For most foundation investigation work, engineers assume sands are drained and have little or no cohesion component. Therefore for sands, the angle of internal friction is needed.

Table 15 tabulates friction angles for specific cohesionless soils. The values show a range from loose to dense material based on the placement void ratio. The values for material No. 8 are from a direct shear test, which is a drained test. The other material values are from triaxial shear tests. It is not known what types of triaxial shear tests were performed, what confining pressure was used, or what failure criteria were used for the friction angle.

Table 14.—Ground-strengthening techniques summarized [20]

Conditions	Technique	Application
Low grades	Compacted sand fill	Minimize structure settlements [1]
Miscellaneous fill		
Shallow	Excavate-backfill	Minimize structure settlement
Deep	Dynamic compaction	Reduce structure settlement [2]
	Sand columns	Reduce structure settlement
Organics		
Shallow	Excavate-backfill	Minimize structure settlement
	Geotextiles	Support low embankments
Deep	Surcharge	Reduce structure settlement
	Geotextiles	Support low embankments
	Sand columns	Reduce structure settlement
Buried	Surcharge	Reduce structure settlement
	Dynamic compaction	Reduce structure settlement
	Compaction grouting	Arrest existing structure settlement
	Sand columns	Reduce structure settlement
Soft clays		
Shallow	Excavate-backfill	Minimize structure settlement
	Geotextiles	Support low embankments
Deep	Surcharge	Reduce structure settlement
	Geotextiles	Support low embankments
	Sand columns	Reduce structure settlement
	Lime columns	Reduce structure settlement
Buried	Surcharge	Reduce structure settlement
	Dynamic compaction	Reduce structure settlement
	Compaction grouting	Arrest existing structure settlement
	Sand columns	Reduce structure settlement
	Lime columns	Reduce structure settlement
Clays, surface	Gravel admixture	Base, subbase, low-quality pavement
	Lime admixture	Stabilize roadway base and subbase
	Freezing	Temporary arrest of settlement
Loose silts		
Shallow	Excavate-backfill	Minimize structure settlement
	Salts admixture	Dust palliative
	Surface compaction	Increase support capacity [3]
Deep	Surcharge	Reduce structure settlement
	Stone columns	Increase support capacity
	Electroosmosis	Increase slope strength temporarily
Buried	Vacuum wellpoints	Improve excavation bottom stability
Loose sands		
Shallow	Surface compaction	Increase support capacity
	Cement admixture	Base, subbase, low-quality pavement
	Bitumen admixture	Base, subbase, low-quality pavement
Deep	VibroflotationjTerra -probe	Increase support capacity
	Dynamic compaction	Increase support capacity
	Stone columns	Increase support capacity
	Wellpoints	Increase stable cut-slope inclination
	Freezing	Temporary stability for excavation

Table 14.—Ground-strengthening techniques summarized [20]

Buried	Penetration grouting	Arrest existing structure settlement
	Freezing	Temporary stability for excavation
Collapsible soils		
Shallow	Excavate-backfill	Minimize structure settlement
Deep	Hydrocompaction	Reduce structure settlement
	Dynamic compaction	Increase support capacity
	Lime stabilization	Arrest building settlement
Liquefiable soils	Dynamic compaction	Increase density
	Stone columns	Pore-pressure relief
Expansive soils	Lime admixtures	Reduce activity in compacted fill
Rock masses		
Fractured	Compaction grouting	Increase strength
	Penetration grouting	Increase strength
	Bolts and cable anchors	Stabilize slopes and concrete dam foundations
	Shotcrete or gunite	Reinforce slopes
	Sub horizontal drains	Stabilize slopes

[1] "Minimize structure settlement" signifies that settlement will be negligible under moderate foundation loads if the technique is applied properly.

[2] "Reduce structural settlement" signifies that after application of the technique, significant settlement, which must be anticipated in the design of the structure, may still occur.

[3] "Increase support capacity" signifies that proper application of the technique will result in an increase in bearing capacity and a decrease in compressibility on an overall basis.

Table 16 tabulates friction angles for sands and silts. Here the terms loose and dense are not defined, and it is not known what types of triaxial shear tests were performed, what confining pressure was used, or what failure criteria were used for the friction angle.

Tables 15 and 16 can be used as a guide of what to expect for values of internal friction for these types of materials.

Figures 19 and 20 graphically show correlations of friction angles and relative density of cohesionless materials. It can be seen from these figures and the previous two tables that, at best, when determining a friction angle for a cohesionless material, a range of values is all that can be expected. Also you will note the dependency of the friction angle on the degree of compaction (relative density) of the sand.

For clays, it is difficult to know the pore water pressures that are generated in rapid loading. Therefore, the engineers want to measure on the cohesion component of the strength. This strength is called the undrained shear strength, S_u. S_u can be estimated from SPT N value, cone penetration, or measured by vane shear, unconfined compression test, or unconsolidated undrained triaxial shear tests on undisturbed samples. The strength in the unconfined compression test is Q_u.

Table 15.—Angle of internal friction of cohesionless soils [13]

No.	General description	Grain shape	D_{10} (mm)	C_u	Loose e	Loose ϕ (°)	Dense e	Dense ϕ (°)
1	Ottawa standard sand	Well rounded	0.56	1.2	0.70	28	0.53	35
2	Sand from St. Peter sandstone	Rounded	0.16	1.7	0.69	31	0.47	37 [1]
3	Beach sand from Plymouth, MA	Rounded	0.18	1.5	0.89	29	-	-
4	Silty sand from Franklin Falls Dam site, NH	Subrounded	0.03	2.1	0.85	33	0.65	37
5	Silty sand from vicinity of John Martin Dam, CO	Subangular to subrounded	0.04	4.1	0.65	36	0.45	40
6	Slightly silty sand from the shoulders of Ft. Peck Dam, MT	Subangular to subrounded	0.13	1.8	0.84	34	0.54	42
7	Screened glacial sand, Manchester, NH	Subangular	0.22	1.4	0.85	33	0.60	43
8 [2]	Sand from beach of hydraulic fill dam, Quabbin Project, MA	Subangular	0.07	2.7	0.81	35	0.54	46
9	Artificial, well-graded mixture of gravel with sands No. 7 and No. 3	Subrounded to subangular	0.16	68	0.41	42	0.12	57
10	Sand for Great Salt Lake fill (dust gritty)	Angular	0.07	4.5	0.82	38	0.53	47
11	Well-graded, compacted crushed rock	Angular	-	-	-	-	0.18	60

[1] The angle of internal friction of the undisturbed St. Peter sandstone is larger than 60°, and its cohesion so small that slight finger pressure or rubbing, or even stiff blowing at a specimen by mouth, will destroy it.
[2] Angle of internal friction measured by direct shear test for No. 8, by triaxial tests for all others.

Cohesion, C, is normally assumed to be ½ the undrained shear strength, S_u, or the unconfined compression test Q_u or U_c.

$$C = \tfrac{1}{2} S_u = \tfrac{1}{2} Q_u$$

Table 17 summarizes clay consistency relationships. In this table, the consistency descriptors are similar to those used in the unified soil classification system, except that system uses five classes of consistency. The undrained strength can be roughly estimated by a simple thumbnail test. Also, handheld pocket

48

Table 16.—Representative values of ϕ_d for sands and silts [21]

Material	Degrees	
	Loose	Dense
Sand, round grains, uniform	27.5	34
Sand, angular grains, well graded	33	45
Sandy gravel	35	50
Silty sand	27-33	30-35
Inorganic silt	27-30	30-34

penetrometers and torvane testers should always be used in the field when exploring clays. Table 17 also shows the expected range of SPT blowcount to be discussed later.

Extreme caution must be taken when using strength values from published tables or figures. The actual material tested, the placement conditions, the testing parameters, and the criteria used to select failure must be known.

3.2.6 Bearing Capacity of Structures

The allowable pressures on structural footing depend on the strength and compressibility of the foundation soils. Footings need to be checked for allowable settlement and for bearing capacity. Most often, the allowable settlement controls the allowable load, yet in some cases, bearing capacity is a concern.

Figures 21 and 22 illustrate how the material will displace under a footing during bearing-capacity failure. Figure 22 is a line footing on sand, and figure 21 is line footing on clay.

Evaluation of bearing capacity depends on the strength of the foundation soils. The general bearing capacity equation is:

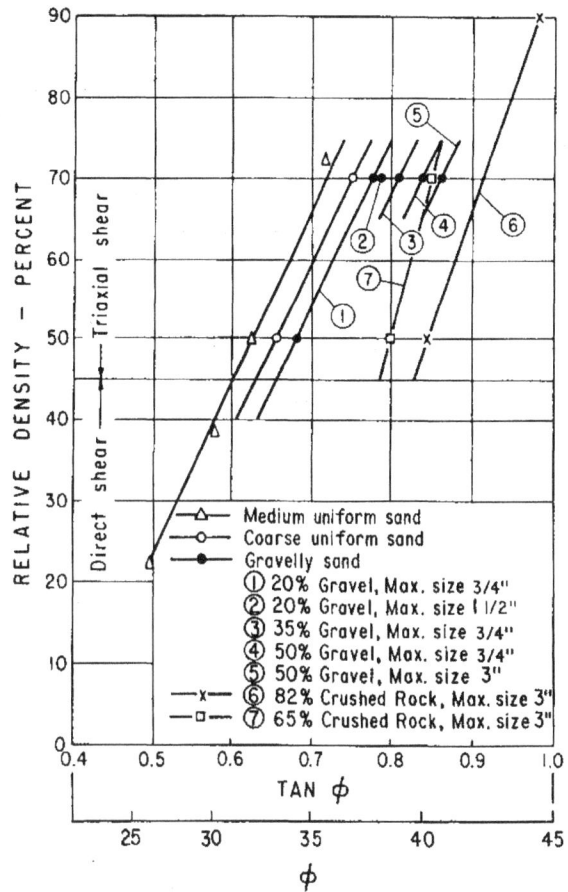

Figure 19.—Effect of relative density on the coefficient of friction, tan ϕ, for coarse grained soils [1].

$$Q_a = K\,C_u\,N_c + K\,\gamma\,N_\gamma B + K\,N_q\,\gamma\,D_f$$

where

Q_a = Allowable footing pressure

$C_u\,N_c$ = Strength component due to clay. C_u is the undrained strength, N_c is the bearing capacity factor.

$\gamma\,N_\gamma\,B$ = Strength component due to sand (friction). N_γ is a function of friction angle.

$N_q\,\gamma\,D_f$ = Strength component due to embedment, surcharge pressure from overlying soil

49

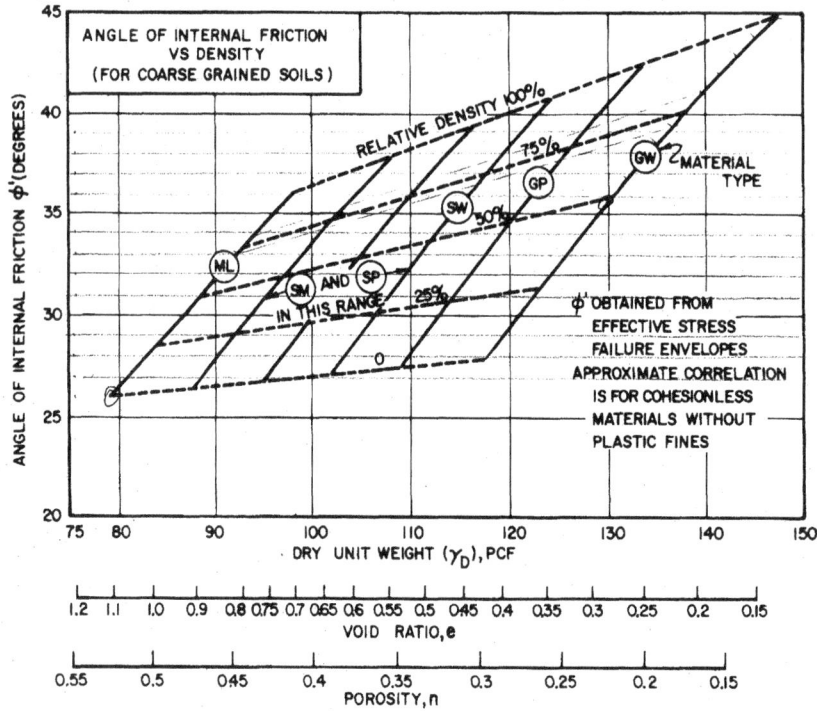

Figure 20.—Correlations between the effective friction angle in triaxial compression and the dry density, relative density, and soil classification [17].

Table 17.—Common properties of clay soils [14]

Consistency	N	Hand test	γ_{sab}[1] g/cm^3	Strength[2] U_c, kg/cm^2
Hard	>30	Difficult to indent	>2.0	>4.0
Very stiff	15-30	Indented by thumbnail	2.08-2.24	2.0-4.0
Stiff	8-15	Indented by thumb	1.92-2.08	1.0-2.0
Medium (firm)	4-8	Molded by strong pressure	1.76-1.92	0.5-1.0
Soft	2-4	Molded by slight pressure	1.60-1.76	0.25-0.5
Very soft	<2	Extrudes between fingers	1.44-1.60	0-0.25

[1] $\gamma_{sat} = \gamma_{dry} + \gamma_w \left(\dfrac{e}{1+e} \right)$

[2] Unconfined compressive strength U_c is usually taken as equal to twice the cohesion c or the undrained shear strength s_u. For the drained strength condition, most clays also have the additional strength parameter ϕ, although for most normally consolidated clays c = 0 [18].

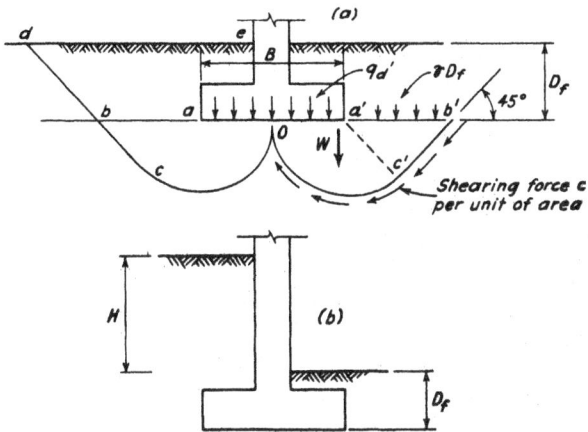

Figure 21.—(a) Cross section through long footing on clay, showing basis for computation of ultimate bearing capacity. (b) Section showing Df for footing with surcharge of different depth on each side [20].

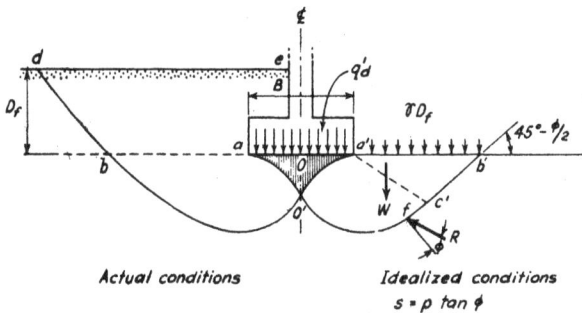

Figure 22.—Cross section through long footing on sand showing (left side) pattern of displacements during bearing-capacity failure, and (right side) idealized conditions assumed for analysis [20].

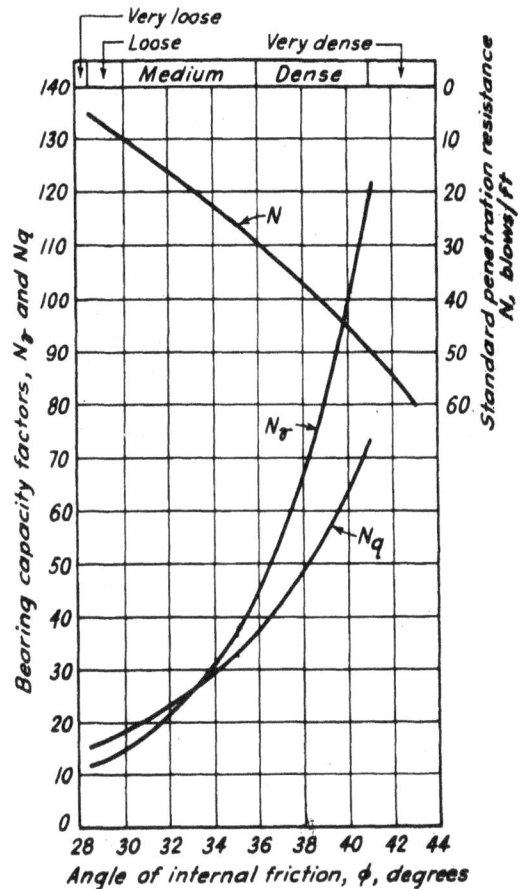

Figure 23.—Curves showing the relationship between bearing-capacity factors and ϕ, as determined by theory, and rough empirical relationship betweeen bearing capacity factors or ϕ and values of standard penetration resistance N [20].

K = constants to account for the footing shape

D_f = depth of footing

B = width of footing (short dimension if retangular)

Figures 23 and 24 show two methods of estimating the bearing capacity. On figure 23, for sands, SPT N value is used to determine the degree of compaction and estimation of bearing capacity factors N_γ and N_q. On figure 24, for clays, allowable pressure is solved through the bearing capacity equation by using the unconfined compressive strength and the ratio of depth to width of the footing.

3.2.7 Settlement

There are two methods to evaluate settlement of structures.

The first step is to estimate the change in pressure due to the structure. Figures 25 and 26 show the stress distribution, pressure bulb, beneath line, and point footings. These diagrams can be used to estimate the pressure

51

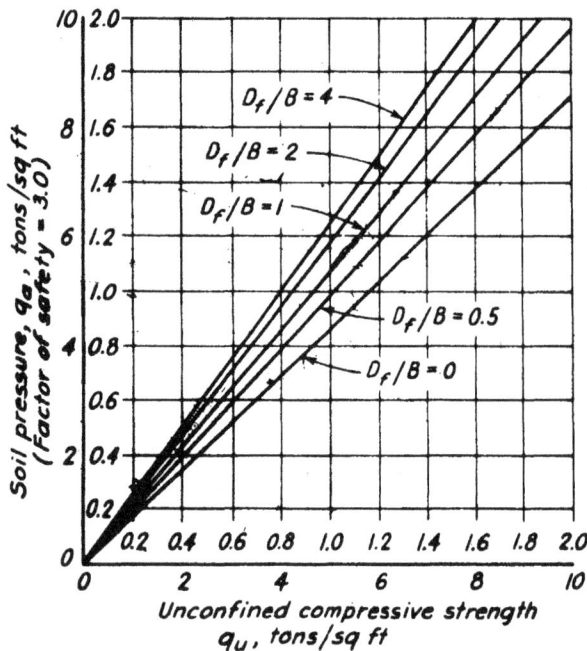

Figure 24.—Net allowable soil pressure for footings on clay and plastic silt, determined for a factor of safety of 3 against bearing capacity failure ($\phi = 0$ conditions). Chart values are for continuous footings (B/L = 0); for rectangular footings, multiply value [20].

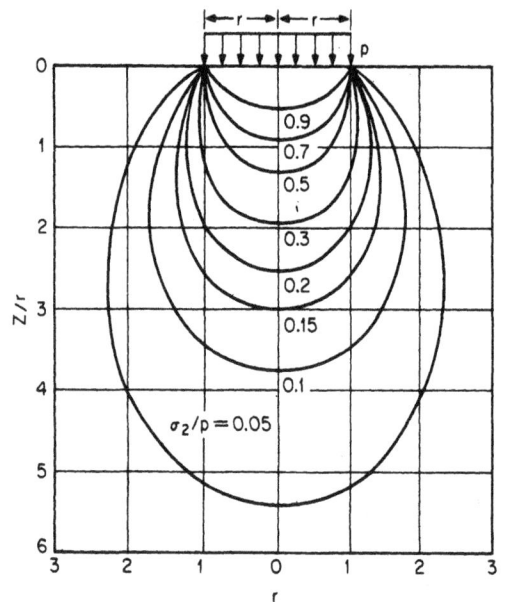

Figure 25.—The "pressure bulb" of stress distribution: contours of vertical normal stress beneath a uniformly loaded circular area on a linear elastic half space [17].

changes beneath a footing to determine the depths for subsurface investigation. Note that 90 percent of the stress concentration occurs at a depth of 1½B to 2B, where B is the short dimension of the footing. The recommendations for drilling depths on figure 6 are based on these types of pressure distribution diagrams. With multiple footings, stresses are superimposed, and the drilling depth may be deeper.

For sands, laboratory consolidation tests are normally not performed, because it is very difficult to obtain undisturbed samples. So for sands, penetration resistance tests are used to estimate soil modulus (stiffness). Typically, the standard penetration test or cone penetration

test is used. In some cases, if the degree of compaction is known, the blow count, N, can be estimated and then the modulus estimated. The most popular method for estimating settlement in sands is called the strain influence factor. Figures 27 and 28 show how settlements in sands are calculated based on a strain influence diagram and estimation of modulus of sand from SPT N value.

Simpler yet, for sands, empirical charts for estimating allowable pressure to limit settlements to less than 1 inch have been developed. In the following section on subsurface investigations, correlations to SPT N values (sec. 3.4.3, figures 37 and 38) can be used to estimate allowable settlement of less than 1 inch for footings.

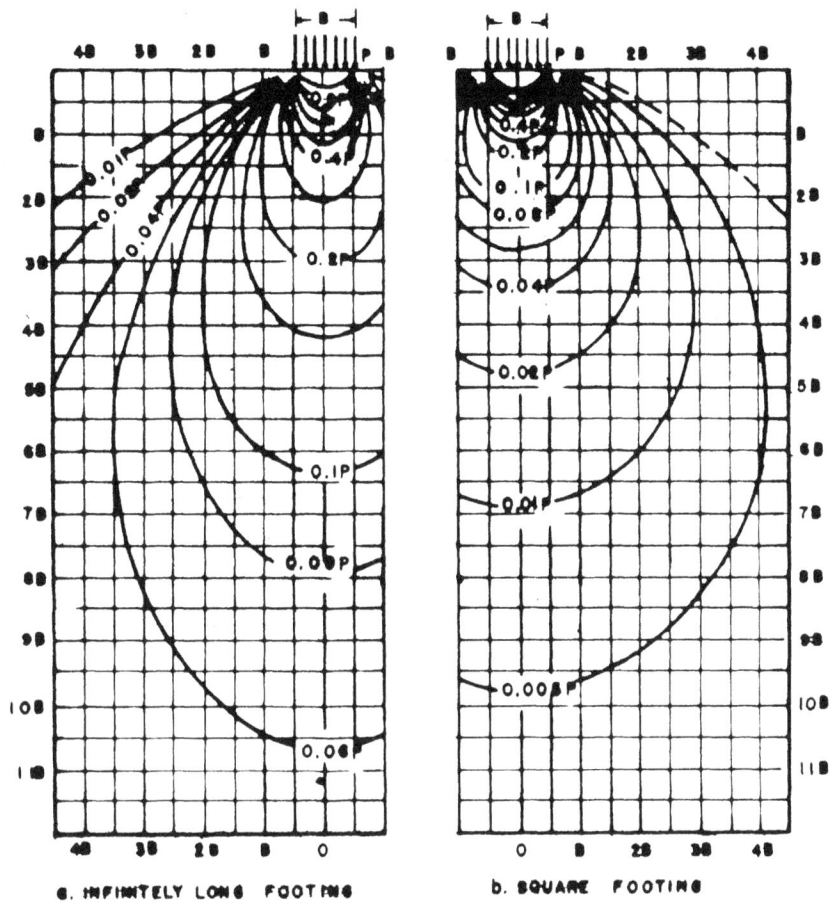

a. INFINITELY LONG FOOTING

b. SQUARE FOOTING

SQUARE FOOTING

GIVEN
  FOOTING SIZE = 20'x 20'
  UNIT PRESSURE P = 2 TSF

FIND
  PROFILE OF STRESS INCREASE
  BENEATH CENTER OF FOOTING
  DUE TO APPLIED LOAD

z (FT)	$\frac{z}{B}$	σ_z TSF		
10	0.5	0.70 X 2	=	1.4
20	1	0.38 X 2	=	0.76
30	1.5	0.19 X 2	=	0.38
40	2.0	0.12 X 2	=	0.24
50	2.5	0.07 X 2	=	0.14
60	3.0	0.05 X 2	=	0.10

B = 20'   P = 2 TSF

Figure 26.—Stress contours and their application [14].

For clays, a one-dimensional consolidation test is performed to measure the compression index C_c, or it may be estimated as in section 3.2.5. Note on figure 14, the consolidation curve, before the virgin consolidation part of the curve, the slope of the compression curve is much flatter. C_c is the slope of the void ratio and pressure curve, and settlements are related to change in specimen height in the compression test. This is called the reloading

53

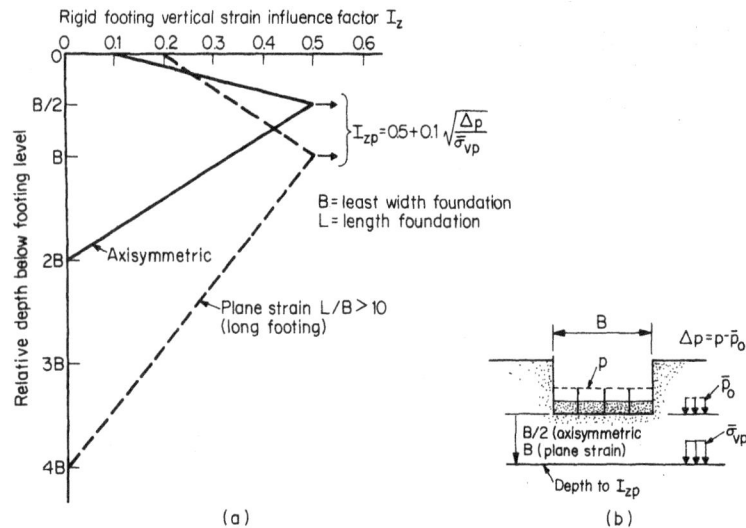

Figure 27.—Modified strain influence factor diagrams for use in Schmertmann method for estimating settlement over sand: (a) modified strain influence factor distributions and (b) explanation of pressure terms in the equation in part (a). In part (a) note that the peak value for the influence factor I_{zp} is found from the equation [22].

portion of the curve, where the clay is reloaded to its maximum past pressure. The slope of this portion of the curve is C_R or recompression index. In cases where surface clays are not saturated (dessicated surface crust) the capillarity of the clay makes it very strong, and if the pressure increases are within the recompression range, settlements will be greatly reduced. However, if the dessicated layer later becomes saturated, major settlements can occur.

Laboratory tests are required for large critical structures or when there are uncertainties in the estimate of C_c for clays. For very small structures, an estimate of C_c may suffice. This estimate requires water contents and/or Atterberg limits for the clay. However, the recompression index cannot be measured from index properties.

Table 18 shows some pre-1930 customary values of allowable soil pressure correlated to material type. These tables were developed prior to the above methods for analyzing settlement. One might be tempted to use these allowable pressures. The tables are fairly reliable for the areas and local geology from which they were developed; however, application to other geologic areas could be dangerous. These tables are useful for understanding general allowable footing loads. For example, compact sands and gravels have the highest allowable footing pressures of 4 to 6 t/ft². Figures 29 and 30 show allowable bearing pressures based on penetration resistance testing. Table 19 is another more recent version of allowable soil pressures, and table 20 is an example from the New York City Building code. If a design uses tables like this, a flag should be raised. The material terms are vague; they should be based on defined soil properties, like classification names and symbols, or index properties.

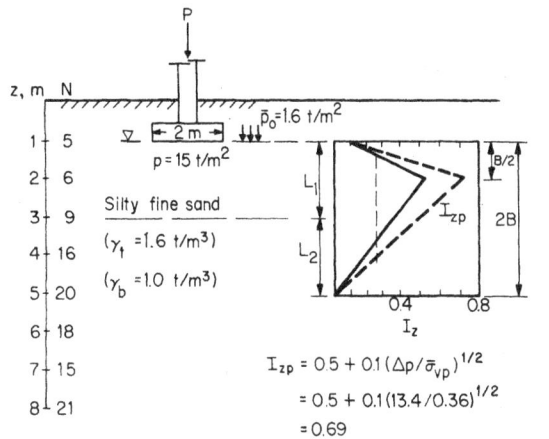

$$I_{zp} = 0.5 + 0.1(\Delta p/\bar{\sigma}_{vp})^{1/2}$$
$$= 0.5 + 0.1(13.4/0.36)^{1/2}$$
$$= 0.69$$

SUMMATION OF STRAIN INFLUENCE						
Layer	Δz, cm	N_1	E_{avg}, kg/cm²	L_z, m*	I_z	$(I_z/E_s) \Delta z$, cm/kg/cm²
1	200	9	100	1	0.73	1.46
2	200	14	140	3	0.24	0.34
						1.80 = $\Sigma(I_z/E_s) \Delta z$

*L_z = depth to midpoint of layer below foundation level.

Find settlements from

$$\rho = C_1 C_2 \, \Delta p (I_z/E) \, \Delta z$$

$$C_1 = 1 - 0.5(\bar{p}_o/\Delta p); \quad \bar{p}_o = 1.6 \times 1.0 = 1.6 \text{ t/m}^2; \quad p = 15 \text{ t/m}^2$$

$$\Delta p = 15 - 1.6 = 13.4 \text{ t/m}^2 = 1.34 \text{ kg/cm}^2$$

$$C_1 = 1 - 0.5(1.6/13.4) = 0.94$$

$$C_2 = 1 + 0.2 \log t_{yr}/0.1, \text{ for } 10 \text{ yr} = 1.4$$

$$\rho = (0.94)(1.4)(1.34)(1.80) = 3.2 \text{ cm; or for } C_2 = 0, \rho = 2.27 \text{ cm}$$

Figure 28.—Computation of settlements in a layered soil deposit using elastic strain concepts for a square footing [22].

3.2.8 Slope Stability

Slopes can be natural or artificial. If the ground surface is not horizontal, a component of gravity will tend to move the material downward. If the component of gravity is large enough and the material's internal shear strength is small enough, a slope failure can occur. There are many types of slope failures, as illustrated in figure 31. These failures differ in speed and the material's water content.

When a slope is checked against potential failure, determination and comparison of the shear stress developed along the most likely rupture surface with the shear strength of the material. The stability analysis of a slope is not an easy task. Evaluation of variables such as stratification and in-place shear strength may prove difficult. Water seepage through the slope and the choice of a potential slip surface add to the complexity of the problem.

One should keep in mind the words of Terzaghi and Peck [18]:

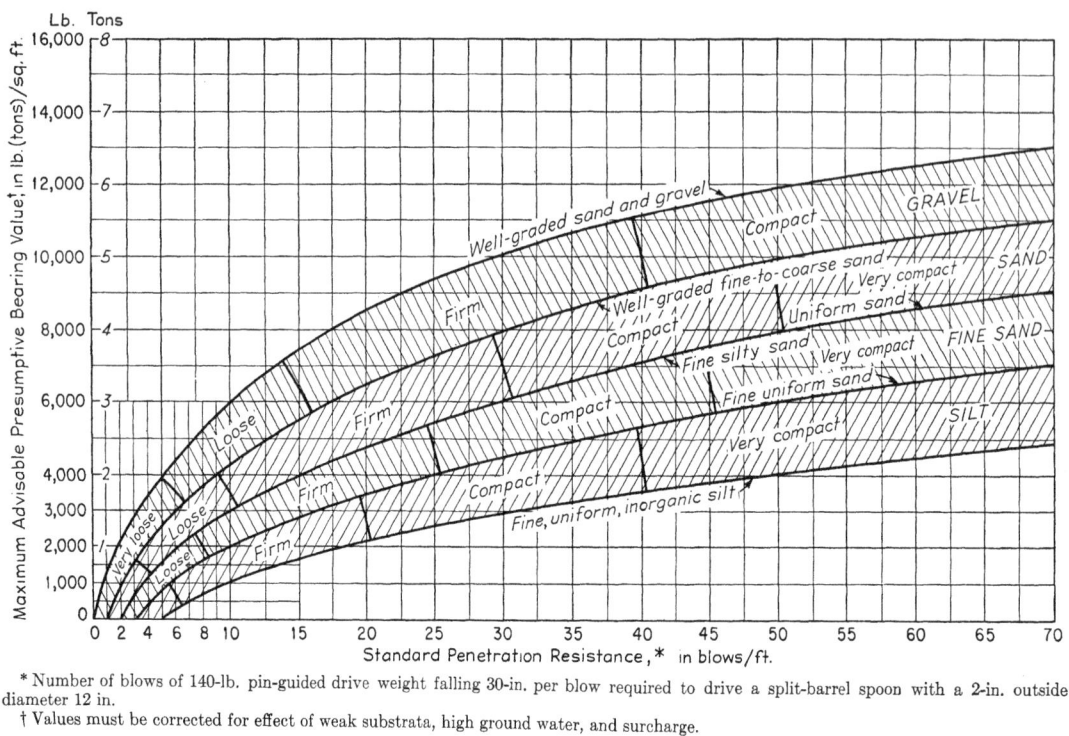

* Number of blows of 140-lb. pin-guided drive weight falling 30-in. per blow required to drive a split-barrel spoon with a 2-in. outside diameter 12 in.

† Values must be corrected for effect of weak substrata, high ground water, and surcharge.

Figure 29.—Presumptive bearing values, granular soils [23].

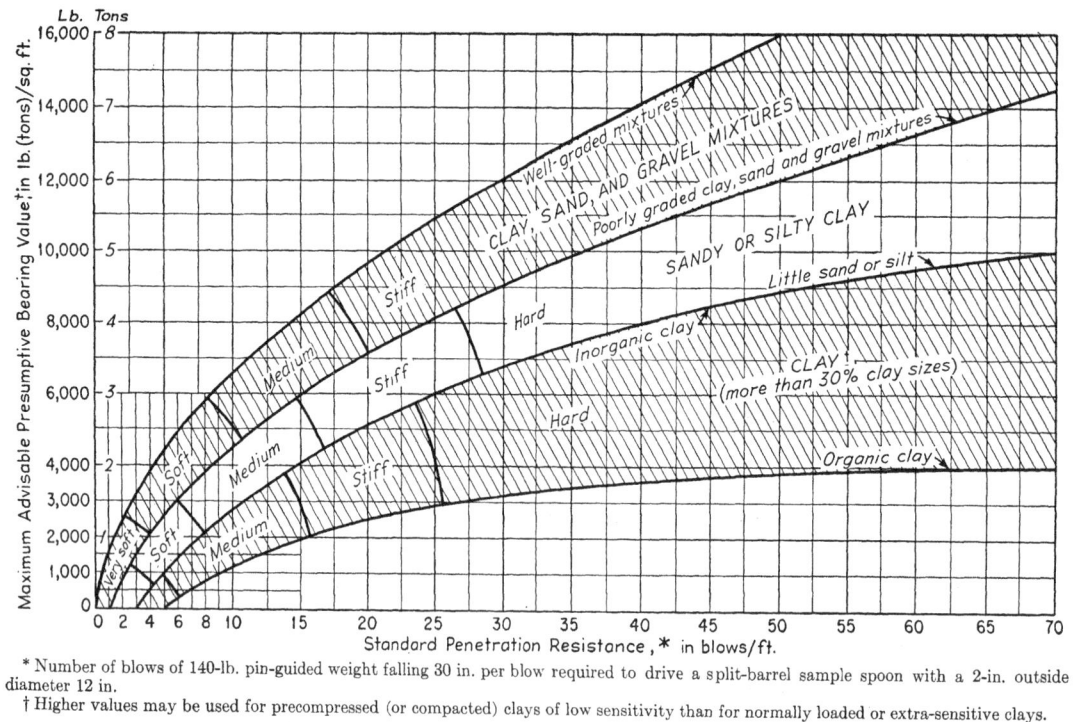

* Number of blows of 140-lb. pin-guided weight falling 30 in. per blow required to drive a split-barrel sample spoon with a 2-in. outside diameter 12 in.

† Higher values may be used for precompressed (or compacted) clays of low sensitivity than for normally loaded or extra-sensitive clays.

Figure 30.—Presumptive bearing values, clays and mixed soils [23].

Table 18.—Soil pressures allowed by various building codes [18]

#	Character of foundation Bed, loads in tons/ft²	Akron, 1920	Atlanta, 1911	Boston, 1926	Cleveland, 1927	Denver, 1927	Louisville, 1923	Minneapolis, 1911	New York, 1922	St. Paul, 1910	Jacksonville, 1922
1	Quicksand or alluvial soil	½	-	-	½	-	-	-	-	-	-
2	Soft or wet clay, at least 15' thick	1	1	-	2	-	-	1	1	-	1
3	Soft clay and wet sand	1½	-	-	1½	-	-	-	-	1	-
4	Sand and clay mixed or in layers	-	2	-	-	-	-	2	2	2	2
5	Firrn clay	-	-	-	-	-	-	-	2	-	-
6	Wet sand	-	-	-	-	-	-	-	2	-	-
7	Fine wet sand	2	-	-	2	-	-	-	-	-	-
8	Soft clay held against displacement	-	-	2	-	-	-	-	-	-	-
9	Clay in thick beds, mod. dry	-	-	-	-	2-4	-	-	-	-	-
10	Dry solid clay	-	-	-	-	-	-	-	-	-	3
11	Loam, clay or fine sand, firm and dry	-	-	-	-	-	2½	3	-	-	-
12	Firm dry loam	2½	2-3	-	-	1-2	-	-	-	-	-
13	Firm dry sand	3	2-3	-	-	2-4	-	-	3	-	3
14	Quicksand when drained	-	-	-	3	-	-	-	-	-	-
15	Hard clay	-	3-4	-	3	-	4	4	-	4	-
16	Fine grained wet sand	-	-	3	-	-	-	-	-	-	-
17	Very firm coarse sand	-	3-4	-	-	4-6	4	4	-	4	4
18	Gravel	-	3-4	-	-	-	4	4	6	4	4
19	Dry hard clay	-	-	-	-	-	-	-	4	-	-
20	Clay in thick beds always dry	4	-	-	-	4-6	-	-	-	-	-
21	Fine dry clay	-	2-3	-	-	-	-	-	-	-	-
22	Fine-grained dry sand	-	-	4	4	-	-	-	-	-	-
23	Compact coarse sand and gravel	-	-	-	-	-	-	-	-	-	4

57

Table 19.—Nominal values for allowable bearing pressure for spread foundations [17]

Type of bearing material	Consistency in place	Allowable bearing pressure, t/ft^2	
		Ordinary range	Recommended value for use
Massive crystalline igneous and metamorphic rock: granite, diorite, basalt, gneiss, thoroughly cemented conglomerate (sound condition allows minor cracks)	Hard, sound rock	60 to 100	80
Foliated metamorphic rock: slate, schist (sound condition allows minor cracks)	Medium hard, sound rock	30 to 40	35
Sedimentary rock: hard cemented shales, siltstone, sandstone, limestone without cavities	Medium hard, sound rock	15 to 25	20
Weathered or broken bedrock of any kind except highly argillaceous rock (shale)	Soft rock	8 to 12	10
Compaction shale or other highly argillaceous rock in sound condition	Soft rock	B to 12	10
Well-graded mixture of fine and coarse-grained soil: glacial till, hardspan, boulder clay (GW-GC, GC, SC)	Very compact	B to 12	10
Gravel, gravel-sand mixtures, boulder-gravel mixtures (GW, GP, SW, SP)	Very compact Medium to compact Loose	6 to 10 4 to 7 2 to 6	7 5 3
Coarse to medium sand, sand with little gravel (SW, SP)	Very compact Medium to compact Loose	4 to 6 2 to 4 1 to 3	4 3 1.5
Fine to medium sand, silty or clayey medium to coarse sand (SW, SM, SC)	Very compact Medium to compact Loose	3 to 5 2 to 4 1 to 2	3 2.5 1.5
Fine sand, silty or clayey medium to fine sand (SP, SM,SC)	Very compact Medium to compact Loose	3 to 5 2 to 4 1 to 2	3 2.5 1.5
Homogeneous inorganic clay, sandy or silty clay (CL, CH)	Very stiff to hard Medium to stiff Soft	3 to 6 1 to 3 0.5 to 1	4 2 0.5
Inorganic silt, sandy or clayey silt, varved silt-clay- fine sand (ML, MH)	Very stiff to hard Medium to stiff Soft	2 to 4 1 to 3 0.5 to 1	3 1.5 0.5

[1] Variations of allowable bearing pressure for size. depth, and arrangement of footings are given in the text.
[2] Compacted fill, placed with control of moisture, density, and lift thickness, has allowable bearing pressure of equivalent natural soil.
[3] Allowable bearing pressure on compressible fine grained soils is generally limited by considerations of overall settlement of structure.
[4] Allowable bearing pressure on organic soils or uncompacted fills is determined by investigation of individual case.
[5] Allowable bearing pressure for rock is not to exceed the unconfined compressive strength.

Table 20.—Allowable bearing pressures for rock and soil [22]

Material	Q_{all}. t/ft^2	Description
Hard, sound rock	60	Crystalline: gneiss, diabase, schist, marble, serpentinite
Medium hard rock	40	Same as hard rock
Intermediate rock	20	Same as hard to medium-hard rock, and cemented sandstones and shales
Soft rock	8	All rocks and uncemented sandstones
Hardpan	12	Groups of GM, GC, and SW, well-cemented and free of lenses of fines and soft rock
Hardpan	8	Groups GM, GC, and SW, poorly cemented with fine-grained matrix or lenses of fines
Gravelly soils	10	Groups GW, GP, GM, GC; compact, well graded
	6	Groups GW, GP, GM, GC; loose, poorly graded
	8	Groups SW, SP, SM; compact, well graded
	4	Groups SW, SP, SM; loose, poorly graded
Sands, coarse-medium	N X 0.1 6 max. 3 min.	Groups SW, SP, SM; with less than 10% of material retained on No. 4 sieve
Sands, fine	N X 0.1 4 max. 2 min. (except for vibratory loads, which require study)	
Hard clay	5	Groups SC, CL, CH; clay requires picking for removal, fresh sample cannot be remolded by finger pressure.
Medium clay	2	Can be removed by spading, can be remolded by substantial finger pressure.
Soft clay		Requires soil testing and analysis. Can be remolded with slight finger pressure.
Silts, dense	3	Groups ML and MH; requires picking for removal.
Silts, medium	1.5	Can be removed by spading.
Silts, loose	-	Requires soil testing and analysis.
Varved silts	2 max.	Higher values permitted when preconsolidated.
Organic soils	1 max.	Untreated
Organic soils	2 max.	Treated by preloading

Slides may occur in almost every conceivable manner, slowly and suddenly, and with or without any apparent provocation. Usually, slides are due to excavation or to undercutting the foot of an existing slope. However, in some instances, they are caused by a gradual disintegration of the structure of the soil, starting at hairline cracks which subdivide the soil into angular fragments. In others, they are caused by an increase of porewater pressure in a few exceptionally permeable layers, or by a shock that liquefies the soil beneath the slope. Because of the extraordinary variety of factors and processes that may lead to slides, the conditions for the stability of slopes usually defy theoretical analysis.

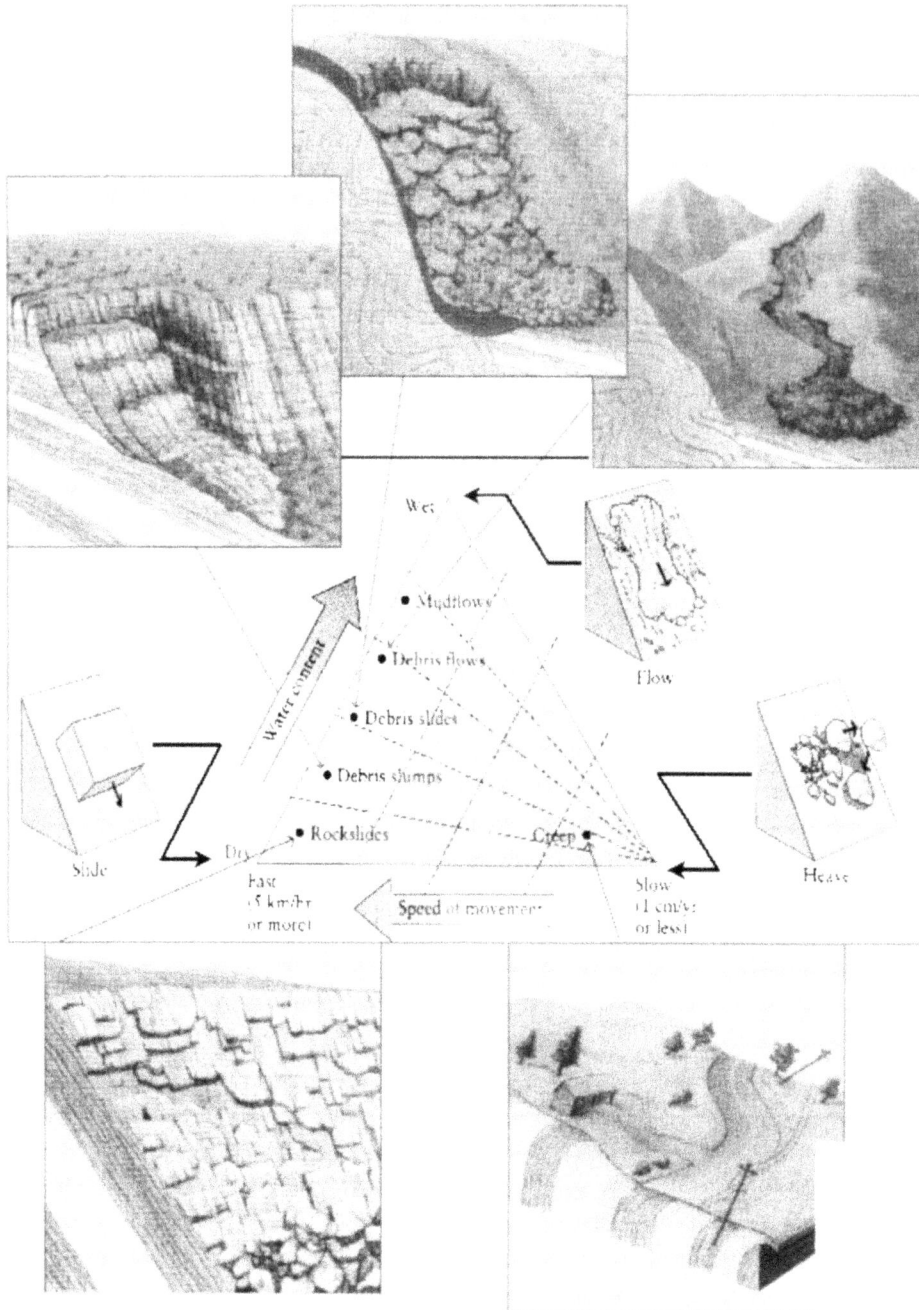

Figure 31.—Styles of slope failure.

Stability of cuts and slopes depends largely upon the shear strength of the materials. For example, sand without cohesion will not stand in a vertical trench in a saturated or dry state, yet clay stands in a vertical trench up to a certain height, depending on the cohesion. This is called the critical vertical height, and it would certainly be dangerous to dig a trench to depths near that critical height. Current Occupational Safety and Health Administration (OSHA) standards must be adhered to on all excavation work. OSHA regulations for trenching were revised in 1990 and contain more stringent requirements, including requirements for responsible engineer or geologist to determine safety and needs for shoring. These OSHA requirements are included in Reclamation health and safety standards.

Problem soils in cuts include weak, soft clays, organic soils, and stiff, fissured clays with weak seams. In some sedimentary rocks, the attitude of the bedding planes may govern the cut slope stability. Sands that will yield ground water will not be stable and run into the excavation.

The assignment of temporary or permanent cut slopes without consideration to site conditions should be discouraged because of the many variables involved in determining stable slopes, such as the type of material, presence of water, depth of cut, and intended use. However, as a guide, from longstanding usage, permanent slopes in soils are commonly excavated at 1½:1 and 2:1 and temporary slopes at 1:1. Some clay soils soften and swell when wetted, and weaken to the extent that much flatter slopes fail. Table 21 provides some typical slope stability problems and preventative measures that can be taken during construction. Additional discussion of slope stability can be found in the *Earth Manual* [1].

An important and simple rule to follow is to observe the success of slopes already existing in the area for indication of expected stability. Also, will any construction activity provoke the failure of the slope by temporally increasing the pore water pressure or deteriorating the strength of the soil?

Generally, slopes can be stabilized by (1) flattening the slope, (2) weighting or anchoring the toe of the slope, (3) unloading the top of the slope, (4) dewatering the zone of slippage by draining, or (5) stabilization of clay by the addition of lime. Table 22 lists possible slope stability remedies.

3.2.9 Seismic Stability

For many sites in the western U.S., seismic stability must be evaluated. If a site is founded on loose alluvial soils, liquefaction can occur. Liquefaction can be evaluated using Reclamation's seismic design standard. Standard penetration, cone penetration, shear wave velocity, and in-place relative density can be used to predict liquefaction potential. Consult the seismic design standard for more information on evaluating seismic stability [24]. Although this document is for evaluating embankment dams, it is useful for other structures. If a structure is founded on liquefiable soils, ground improvement or pile foundations might be required. Other seismic problems include lateral spreading, and fault offsets. Lateral spreading can occur with ground sloping of 2 to 3 percent. Settlement can also occur due to liquefaction. Typical settlements are 3 to 5 inches, but settlement of up to 12 inches has occurred. Geologic reports should indicate potential faulting and slope stability/spreading problems.

3.2.10 Backfill Material

Personnel engaged in the operation of a water system are required to maintain, repair, and occasionally engage in relocation of existing structures, or construct new structures. The majority of this work involves the use of soil as a construction material. Therefore, they should have a working knowledge of the properties of soils, which will aid in the proper selection for a

Table 21.—Slope failure forms: typical preventive and remedial measures [14]

Failure form	Prevention during construction	Remedial measures
Rock fall	Base erosion protection Controlled blasting excavation Rock bolts and straps, or cables Concrete supports, large masses Remove loose blocks Shotcrete weak strata	Permit fal!, clean roadway Rock bolts and straps Concrete supports Remove loose blocks Impact walls
Soil fall	Base erosion protection	Retention
Planar rock slide	Small volume: remove or bolt Moderate volume: provide stable inclination or bolt to retain Large volume: install internal drainage or relocate to avoid	Permit slide, clean roadway Remove to stable inclination or bolt Install internal drainage or relocate to avoid
Rotational rock slide	Provide stable inclination and surface drainage system Install internal drainage	Remove to stable inclination Provide surface drainage Install internal drains
Planar (debris) slides	Provide stable inclination and surface drainage control Retention for small to moderate volumes Large volumes: relocate	Allow failure and clean roadway Use preventive measures
Rotational soil slides	Provide stable inclination and surface drainage control, or retain	Permit failure, clean roadway Remove to stable inclination, provide surface drainage, or retain Subhorizontal drains for large volumes
Failure by lateral spreading	Small scale: retain Large scale: avoid and relocate, prevention difficult	Small scale: retain Large scale: avoid
Debris avalanche	Prediction and prevention difficult Treat as debris slide Avoid high-hazard areas	Permit failure, clean roadway; eventually self-correcting Otherwise relocate Small scale: retain or remove
Flows	Prediction and prevention difficult Avoid susceptible areas	Small scale: remove Large scale: relocate

specific purpose. Of equal importance to proper selection is to realize and understand that the successful use of a soil depends upon proper processing—that is, to increase or decrease the moisture content as required and thoroughly mix to form a uniform homogeneous mixture—and finally, proper placement and compaction.

Table 22.—Summary of slope treatment methods for stabilization [22]

Treatment	Conditions	General purpose (preventive or remedial)
CHANGE SLOPE GEOMETRY		
Reduce height	Rotational slides	Prevent/treat during early stages
Reduce inclination	All soil/rock	Prevent/treat during early stages
Add weight to toe	Soils	Treat during early stages
CONTROL SURFACE WATER		
Vegetation	Soils	Prevent
Seal cracks	Soil/rock	Prevent/treat during early stages
Drainage system	Soil/decomposing rock	Prevent/treat during early stages
CONTROL INTERNAL SEEPAGE		
Deep wells	Rock masses	Temporary treatment
Vertical gravity drains	Soil/rock	Prevent/treat during early stages
Subhorizontal drains	Soil/rock	Prevent/treat-early to intermediate stages
Galleries	Rock/strong soils	Prevent/treat during early stages
Relief wells or toe trenches	Soils	Treat during early stages
Interceptor trench drains	Soils [cuts/fills]	Prevent/treat during early stages
Blanket drains	Soils [fills]	Prevent
Electroosmosis *	Soils [silts]	Prevent/treat during early stages: temporarily
Chemicals *	Soils (clays]	Prevent/treat during early stages
RETENTION		
Concrete pedestals	Rock overhang	Prevent
Rock bolts	Jointed or sheared rock	Prevent/treat sliding slabs
Concrete straps and bolts	Heavily jointed or soft rock	Prevent
Cable anchors	Dipping rock beds	Prevent/treat early stages

63

Table 22.—Summary of slope treatment methods for stabilization [22]

Treatment	Conditions	General purpose (preventive or remedial)
Wire meshes	Steep rock slopes	Contain falls
Concrete impact walls	Moderate slopes	Contain sliding or rolling blocks
Shotcrete	Soft or jointed rock	Prevent
Rock-filled buttress	Strong soils/soft rock	Prevent/treat during early stages
Gabion wall	Strong soils/soft rock	Prevent/treat during early stages
Crib wall	Moderately strong soils	Prevent
Reinforced earth wall	Soils/decomposing rock	Prevent
Concrete gravity walls	Soils to rock	Prevent
Anchored concrete curtain walls	Soils/decomposing rock	Prevent/treat—early to intermediate stages
Bored or root piles	Soils/decomposing rock	Prevent/treat—early stages

* Provides strength increase

3.3 Surface Investigations

Further surface investigations include geologic mapping and the use of geophysics. Surface geologic mapping is required for any investigation. Preliminary geologic maps should be generated prior to field investigations to aid in the selection and location of investigations. Final surface geologic maps in more detail are required for construction specifications.

3.3.1 Surface Geophysics

Surface geophysical investigations should be considered, especially for long line structures.

Surface geophysics using resistivity (ASTM G-57 [25, 16]) are often required to determine corrosion potential for structures. Shear wave velocity measurements are useful for line structures, especially when rock excavation may be encountered. Figure 32 shows how shear wave velocity can be used to determine rock ripability for a D9N dozer. Other geophysical methods may be of benefit, as they generally can reduce the amount of borings by filling in gaps in the exploration. For more information on geophysics, consult the *Earth Manual* [1] and Engineering Geology Manuals [3, 4].

64

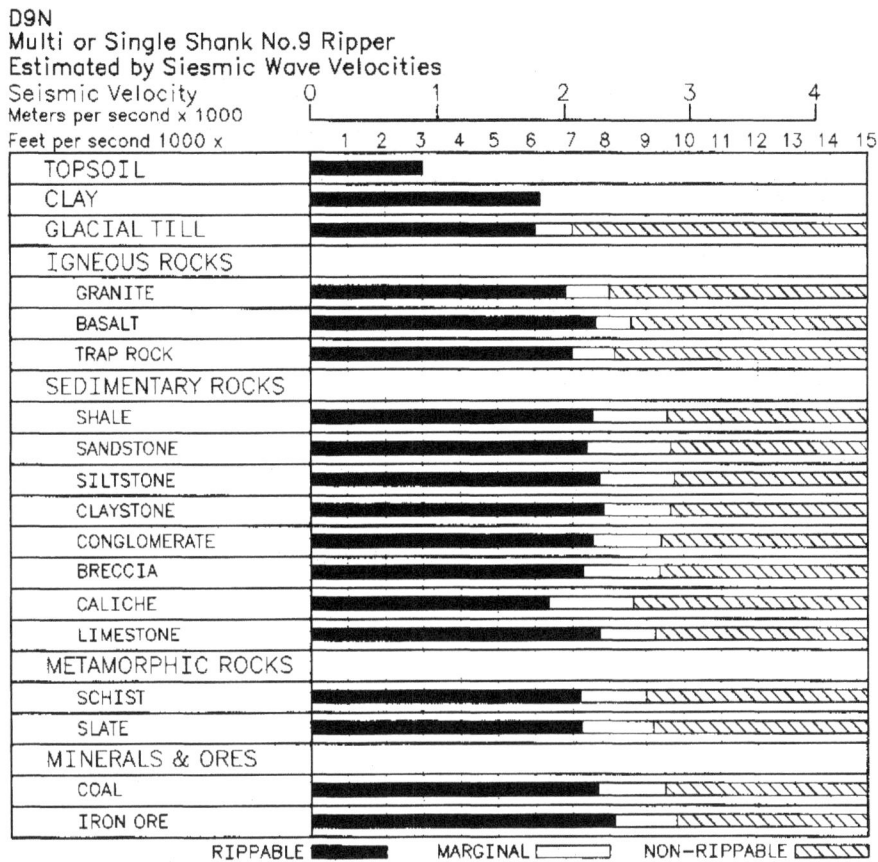

Figure 32.—Rock rippability as related to seismic P-wave velocities (courtesy of Caterpillar Tractor Co.) [1].

3.4 Subsurface Exploration

Many of the test procedures described in this section are discussed in chapter 2 of the *Earth Manual*, Part 1 [1]. Bureau of Reclamation procedures are denoted by test procedure numbers, in the form of USBR XXXX, where XXXX is a four-digit number. The reader should refer to the *Earth Manual* [2] test procedure manual for more information, if these procedures are used on a specific project. Table 23 lists the testing sample sizes for the tests, as listed in table 8. This table can be used to determine the extent of the exploration required.

Table 24 lists exploration methods that are discussed later. The table briefly summarizes the exploration method application and limitations. It also lists the tests, by index numbers, that apply to the exploration method. The last column lists the cost of the exploration method relative to the cost for a test pit.

Using the costs for the exploration from table 24 and the testing costs in table 21, an estimated cost to obtain a soil design parameter can be determined.

Table 23. — Testing sample size

Test index	Test	Maximum particle size	Minimum sample size / dry mass			Remarks[4]
			Undisturbed[1]	Disturbed[2]	Reconstituted[3] test specimen	
1	Visual classification	3-inch	50 g	50 g	50 g	
2	Inplace density					
3	Nuclear moisture-density					
4	Laboratory classification Gradation	minus No. 4	100 g	5 lbm	100 g	
		⅜-inch	200 g	5 lbm	200 g	
		¾-inch	1100 g	15 lbm	1100 g	
		1½-inch	20 lbm	100 lbm	20 lbm	
		3-inch	150 lbm	200 lbm	150 lbm	
	Atterberg limits	minus No. 40	150 g	5 lbm	150 g	
5	Specific gravity	minus No. 4	100 g	5 lbm	100 g	
		¾-inch	3 kg	30 lbm	3 kg	
		1½-inch	5 kg	50 lbm	5 kg	
		3-inch	18 kg	200 lbm	18 kg	
6	Laboratory compaction standard	No. 4 fraction	50 lbm	200 lbm	50 lbm	4 & 5 can be obtained from disturbed sample
7	Laboratory compaction large scale	3-inch	225 lbm	350 lbm	225 lbm	1 specimen / 4 & 5 can be obtained from disturbed sample
8	Relative density coarse material	¾-inch	75 lbm	200 lbm	75 lbm	4 can be obtained from disturbed sample
		3-inch	225 lbm	350 lbm	225 lbm	
9	Dispersive analysis crumb	minus No. 4	1.5-inch dia. length, 2-inch	5 lbm	400 g	4 can be obtained from disturbed sample, or undisturbed trimmings
	Pinhole				500 g	
	Double hydrometer				800 g	

66

Table 23.—Testing sample size

Test index	Test	Maximum particle size	Minimum sample size / dry mass			Remarks[4]
			Undisturbed[1]	Disturbed[2]	Reconstituted[3] test specimen	
10	Well permeameter					
11	Field permeability					
12	Permeability Falling head Constant head	No. 4 3-inch		50 lbm 350 lbm	15 lbm 225 lbm	4 & 5 can be obtained from disturbed sample
13	Back pressure permeability	No. 4	3.25-inch dia. length, 4-inch	240 lbm	16 lbm	4 & 5 can be obtained from disturbed sample, or undisturbed trimmings
14	Flow pump permeability	No. 4	6.0-inch dia. length, 5-inch	16 lbm	2 lbm	4 & 5 can be obtained from disturbed sample, or undisturbed trimmings
15	Filter	3/8-inch base 3-inch filter	50 lbm 100 lbm	50 lbm 100 lbm	50 lbm 100 lbm	4 can be obtained from disturbed sample
16	One-dimensional consolidation	No. 4	4.25-inch dia. length, 2-inch	15 lbm	1.5 lbm	4 & 5 can be obtained from disturbed sample, or undisturbed trimmings
17	One-dimensional expansion	No. 4	4.25-inch dia. length, 2-inch	15 lbm	1.5 lbm	4 & 5 can be obtained from disturbed sample, or undisturbed trimmings
18	One-dimensional uplift	No. 4	4.25-inch dia. length, 2-inch	15 lbm	1.5 lbm	4 & 5 can be obtained from disturbed sample, or undisturbed trimmings
19	Angle of Repose	No. 4 3/8-inch 3/4-inch 1½-inch 3-inch		60 lbm 150 lbm 300 lbm 500 lbm 500 lbm	60 lbm 150 lbm 300 lbm 500 lbm 500 lbm	4 can be obtained from disturbed sample

Table 23. — Testing sample size

Test index	Test	Maximum particle size	Minimum sample size / dry mass			Remarks [4]
			Undisturbed [1]	Disturbed [2]	Reconstituted [3] test specimen	
20	Unconfined compressive strength	No. 4	2-inch dia. length, 6-inch	20 lbm	2 lbm	4 & 5 can be obtained from disturbed sample, or undisturbed, trimmings
		¾-inch	6-inch dia. length, 15-inch	100 lbm	45 lbm	
		1-½-inch	19-inch dia. length, 24-inch	200 lbm	150 lbm	
21	Field vane shear					
22	Laboratory vane shear	No. 4	3.5-inch dia. length, 4.0-inch			4 & 5 can be obtained from undisturbed trimmings
23	Direct shear fine grain material	No. 4	2-inch sq. 3-inch dia. length, 2-inch	10 lbm	4 lbm	4 specimens 4 & 5 can be obtained from disturbed sample, or undisturbed trimmings
24	Direct shear coarse grain material	⅜-inch	6-inch sq. (9-inch dia.) length, 3-inch		15 lbm	1 specimen
		1-½-inch	9 - in sq. (15-inch dia.) length, 4-inch		30 lbm	4 & 5 can be obtained from disturbed sample, or undisturbed trimmings
		1-½-inch	12-inch sq. (20-inch dia.) length, 4-inch		75 lbm	
25	Repeated direct shear	No. 4	4-inch sq. (6-inch dia.) length, 3-inch	20 lbm	8 lbm	4 specimens 4 & 5 can be obtained from disturbed sample, or undisturbed trimmings
26	Rotational shear	No. 200	300 g	300 g	300 g	1 specimen
27	Simple shear					

Table 23.—Testing sample size

Test index	Test	Maximum particle size	Minimum sample size / dry mass			Remarks [4]
			Undisturbed [1]	Disturbed [2]	Reconstituted [3] test specimen	
28	Initial capillary pressure	No. 4	2-inch dia. length, 5-inch	16 lbm	2 lbm	4 & 5 can be obtained from disturbed sample, or undisturbed trimmings
29	Coefficient of earth pressure at rest, K_0	No. 4	2-inch dia. length, 6-inch	20 lbm	2 lbm	1 specimen
30	Unconsolidated-undrained triaxial shear, UU	¾-inch dia.	length, 15-inch 19-inch dia.	100 lbm	95 lbm	1 specimen (3 required per suite)
31	Consolidated-undrained triaxial shear, CU	1½-inch	length 24-inch	200 lbm	150 lbm	
32	Consolidated drained triaxial shear, CD					

[1] Undisturbed: length is the minimum for a continuous specimen
[2] Disturbed: the minimum sample to get a representative specimen
[3] Reconstituted: minimum amount of material for a specimen
[4] Remarks: number represents the other tests required

69

Table 24.—Exploration methods.

Exploration method	Application	Limitations	Test index application	Cost relative to test pitting
Trenching	Soil strata identification. Groundwater seepage and level. Recovery of disturbed or undisturbed samples above groundwater, and in situ density tests. Examination of fault zones.	Usually limited in depth by water table, rock depth, or reach of equipment. Must meet OSHA requirements.	All	2 to 5 x
Test pits	Soil strata identification. Groundwater seepage and level. Recovery of disturbed or undisturbed samples above groundwater, and in situ density tests. Examination of fault zones.	Usually limited in depth by water table, rock depth, or reach of equipment. Must meet OSHA requirements. Usually more expensive than trenching.	All	1
Hand auger	Disturbed samples for classification, and index properties testing.	Disturbed Sample. Sample may not represent a large area. Depth about 20 feet. Limited to ground water table. Slow in hard soils.	1, 4, 5	1
Power auger	Disturbed samples for classification, and index properties testing. Up to 16-inch diameter samples. Normally used in cohesive soils with adequate strength to prevent open hole collapse.	Small sample size. Hole collapses when auger withdrawn from weak cohesive or cohesionless granular soils, thereby limiting depth, usually to near water table.	1, 4, 5, 10, 11	1
Bucket auger	Drill large-diameter (48-inches) holes for disturbed samples and soil strata examination in cohesive soils where hole remains open.	Disturbed samples. Depth limited by groundwater and rock conditions. Not suitable in cohesionless soils, soft clays, or organic soils.	1, 4, 5	3 to 7 x
Thinwall push tube	Undisturbed samples up to 5-inch diameter in soft to firm clays and silts.	Will not penetrate compact sands, stiff clays, and other strong soils. Will not retrieve sands. Can be over pushed.	All limited by sample size	2 to 5 x
Hollow stem auger	Continuous undisturbed samples (with liner) up to 6-inch diameter. Good for Loessial soils to obtain inplace density.	Penetration in strong soils to significant depths or through gravel layers difficult, and not possible through boulders, and rock.	All limited by sample size	5 to 7 x

70

Table 24.—Exploration methods.

Exploration method	Application	Limitations	Test index application	Cost relative to test pitting
Pitcher sampler	Undisturbed samples up to 6-inch diameter in strong cohesive soils such as residual soils, glacial till, soft rock. Superior in alternating soft to hard layers. Can be used in firm clays.	Not suitable in clean granular soils.	All limited by sample size	6 to 8 x
Dension sampler	Undisturbed samples up to 6-inch diameter in strong cohesive soils such as residual soils, glacial till, soft rock.	Not suitable in clean granular soils, and soft to firm clays.	All limited by sample size	6 to 8 x
Standard penetration Test	Recovery of small disturbed samples and determination of soil profile	Penetration limited to soils and soft rocks. Not suitable for boulders and hard rocks.	1	2 to 5 x
Cone penetration testing	Continuous penetration resistance including side friction and point resistance for all but very strong soils. Can be used below water table.	No samples recovered.		1 to 2 x

3.4.1 Test Pits and Dozer Trenches

Test pits and trenches are an excellent form of accessible investigation because large exposures of in-place conditions can be observed. In addition to identifying soil types, the actual soil structure can be observed and recorded. In the past, cribbed test pits were excavated by hand to depths up to 50 to 75 feet. However, today with more stringent safety requirements, test pit depths are restricted, and their value has been diminished.

OSHA regulations require that pits deeper than 5 feet have a stability evaluation by a responsible person. Often, personnel are reluctant to make decisions regarding stability and often, pits are then limited to depths of 5 to 10 feet. This depth is often not sufficient to provide information at depth as required.

Registered geologists and geotechnical engineers should be able to make reasonable decisions regarding stability but should err on the side of safety by using sloping and benching and/or shoring as required. Often pits excavated in dry ground are stable, but after rainfall or saturation, they may become unstable. Test pits excavated below the water table tend to cave and do not provide valuable information. Test pits where only the shovel spoils are observed, are also of little value because of mixing, and important structural information, such as weak seams, may be missed. In order for test pits to yield the most information, the faces of the pits should be cleaned, and the in-place structure should be observed and recorded.

Test pits (USBR 7000) are logged in accordance with Reclamation's *Soil Classification Handbook* [21], and soils are classified according to the Unified Soil Classification System, USBR 5005, and USBR 5000. This allows for uniform presentation of data to designers and contractors. To save investigation costs, visual classification of soils is often sufficient for design of small structures. However, at any new site the investigators should request a certain amount of laboratory data to substantiate visual classifications, and to help the investigator calibrate his judgments on soil particle size and consistency. For investigation of pipe (sec. 1.2.2), inplace density and degree of compaction evaluation is required in a test pit at approximate pipe elevation. Sketches of the various soil strata encountered in test pits are very helpful for designers and contractors.

Accessible test pits allow for direct sampling of the materials. It is important to procure natural moisture content data (USBR 5300) of fine-grained soils, because engineering properties can be estimated from Atterberg limits and moisture contents of clays. Simple tests, such as those with pocket penetrometers or Torvanes (USBR 5770), can be run on the in-place material for estimates of strength.

It is recommended that in-place density be measured in all test pits and the degree of compaction determined by laboratory compaction tests performed on the same material where in-place density test was taken. This allows for direct measure of consistency and refined estimation of engineering properties. In-place density with sand cone and laboratory compaction tests are inexpensive and take little extra time. When pipe construction and borrow studies are being investigated, in-place density tests (USBR 7205) and compaction tests (USBR 5500 or USBR 7240)

are required to be performed on the spring line elevation of the pipe.

Composite samples of the trench walls can be taken to anticipate mixing for borrow materials (USBR 7000). This is especially true if soils are to be blended for construction. If critical soils are identified for laboratory testing, high quality block samples can be taken (USBR 7100). The locations of all tests should be noted clearly on the test pit logs.

In some geologic formations such as alluvial deposits, it is helpful to extend test pits into long trenches so changes in materials can be observed.

3.4.2 Hand Auger

Small auger holes cannot be logged and sampled as accurately as an open trench or a test pit because they are inaccessible for visual inspection of the total profile and for selecting representative strata. Procedures for augering and sampling are discussed in USBR 7010. Small hand augers can be used to collect samples adequate for soil classification and, possibly, for index property testing if the sample size requirements are met [1, page 143].

The hole created by the auger can be used to determine the average coefficient of permeability for soil in its natural condition by performing the field permeability testing by the well permeameter method (USBR 7300). Borehole smear should be removed by roughing with a wire brush.

Figure 33 illustrates how the hand auger can be used to separate soil types as the augering proceeds.

Figure 33.—Hand auger sampling [1].

SAMPLE No 1
FROM HERE
(Similar soil)

Piles are separated when
significantly different materials
are encountered.

SAMPLE No 2
FROM HERE
(Similar soil)

3.4.3 Standard Penetration Resistance Testing

The penetration resistance (PR) test is also commonly called the "standard penetration test, SPT." However, the test is by no means standardized in the U.S., because a wide array of hammers, drilling methods, and samplers are allowed. Over the last 10 years, there have been significant improvements in the test, especially the use of automatic hammer systems. USBR 7015 gives Reclamation procedures for performing PR tests. PR testing is one of the most commonly performed tests in the U.S. for performing foundation investigations for a wide array of applications.

Penetration resistance tests provide a soil sample for identification and for limited index property testing. The classification information is used to develop site stratigraphy and to identify zones where further, more detailed investigations may be required. Many widely published correlations, as well as local correlations, are available that relate penetration resistance to engineering behavior of earthwork and foundations. Local geotechnical testing firms develop local correlations. The following table and figures correlate some PR test data (blow counts, N) to various soil parameters. When using the data for the correlations, the method of PR testing and equipment used should be taken into consideration. PR testing is subject to many types of operator or

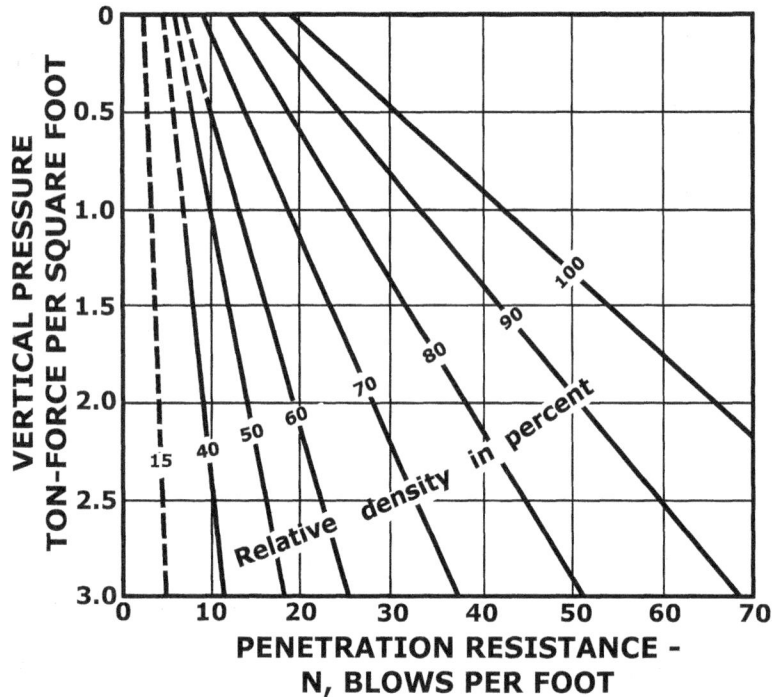

Figure 34.—Correlations between relative density and standard penetration resistance.

mechanical errors. Users should read the Dam Safety Office report on performing SPT prior to testing [26].

Table 25 lists the well known and universally accepted correlation between the SPT blow counts, N, and relative density of cohesionless soils proposed by Terzaghi and Peck [18]. Using a range of relative densities, material compactness terms were defined. Using the relative density, one can estimate parameters such as the friction angle of the material. This chart is generally for shallow ground conditions of about 1 t/ft^2 effective stress. It does not account for confining pressure.

Figure 34 shows the correlation between the SPT blow count, N, and vertical effective stress to relative density of clean sands. At a constant relative density, SPT N value increases with

Table 25.—Correlations for cohesionless soils between compactness, D_R, and N [18]

Compactness	Relative density D_R*	N (SPT)
Very loose	<0.15	<4
Loose	0.15-0.35	4-10
Medium dense	0.35-0.65	10-30
Dense (compact)	0.65-0.85	30-50
Very dense	0.85-1.0	>50

depth as the effective overburden pressure increases. This correlation is for clean quartz sands. and is not applicable to sands with over 10 percent fines It was developed from chamber tests performed by Reclamation in the 1950s and confirmed by additional chamber tests by the U.S. Army Corps of Engineers.

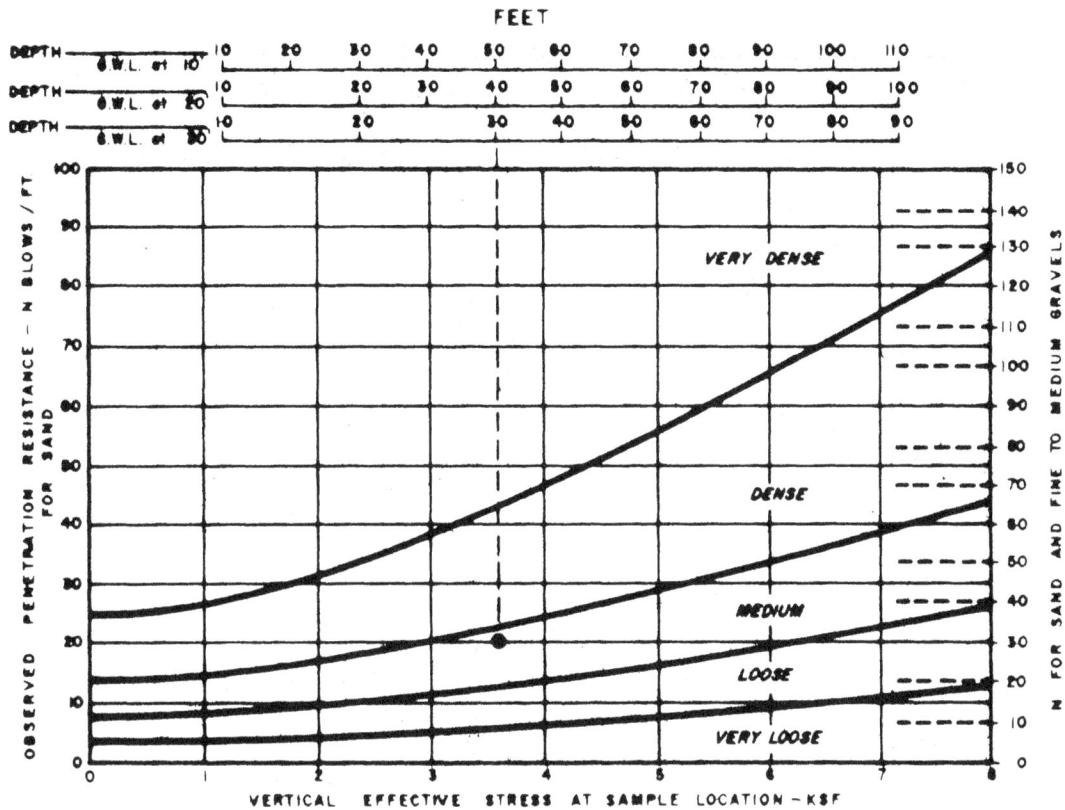

Example:

Blow count in sand at a depth of 40 ft = 20
Depth of Groundwater Table = 20 ft
Compactness ∼ medium

Figure 35.—Estimated compactness of sand from standard penetration test [17].

This chart finds widespread use in geotechnical practice.

Figure 35 shows a correlation between the SPT blow count, N, to relative density and vertical effective stress for sands published in NAVFAC DM-7.2 [17]. The chart also purports to scale for sand and fine to medium gravels. However, the basis is unknown, and it is believed to be just an estimate. There are no chamber tests for gravelly soils. When gravels are encountered, the SPT is generally unreliable, because the blow count is elevated. Note, the ranges of

relative densities for the compactness terms are slightly different than the ones in table 25.

Figure 36 tabulates and graphically shows a correlation between SPT blow counts, N, and unconfined compressive strength of cohesive soils [17]. Sowers proposed lines for different levels of plasticity in cohesive soils. Actually, the N value is a poor predictor of undrained strength. Tabulations of data have shown a wide range in S_u. If S_u needs to be predicted more accurately, cone penetration tests, vane shear tests, or laboratory testing of undisturbed samples can be used.

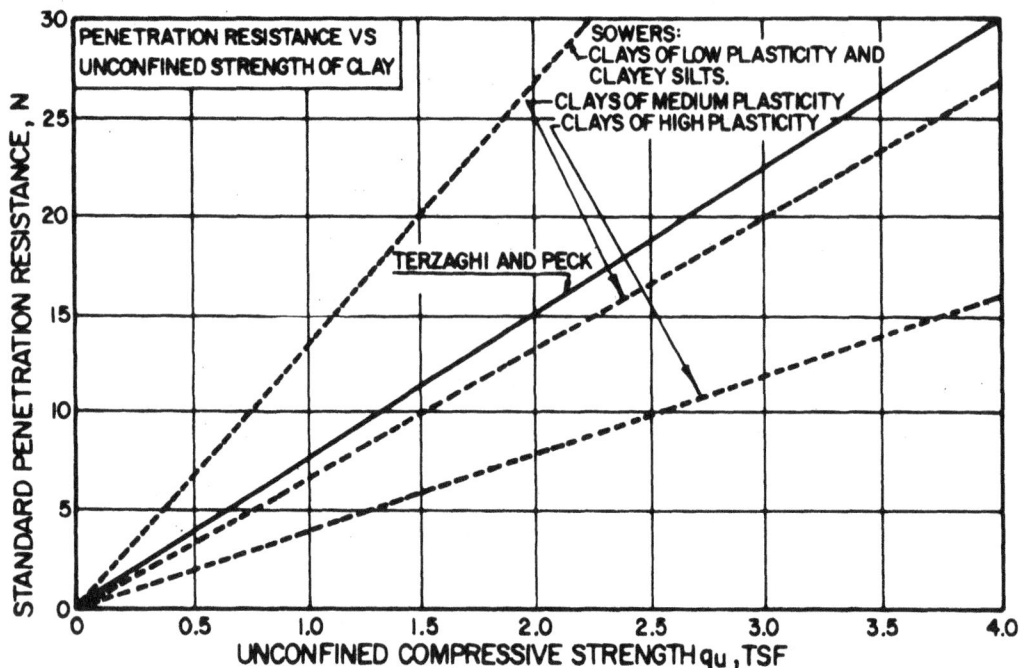

Figure 36.—Correlations of SPT N values with U_c for cohesive soils of varying plasticities [17].

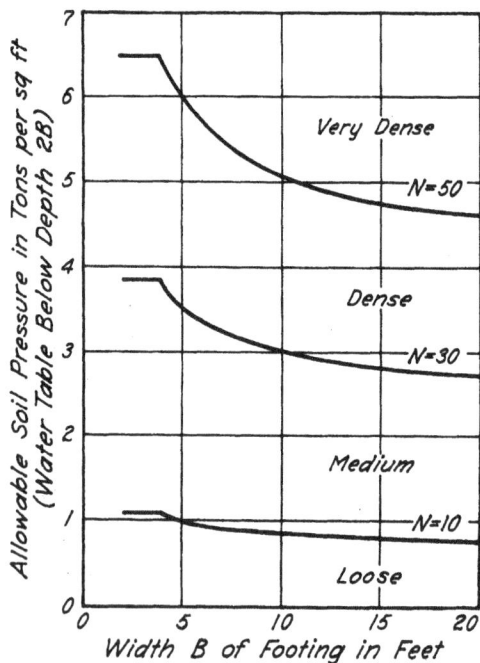

Figure 37.—Chart for estimating allowable soil pressure for footings on sand on the basis of results of standard penetration test [18].

Figure 37 shows a chart for estimating allowable soil pressure for footings on sand on the basis of results from SPT tests. This chart is for very shallow footing, since the depth of the footing is not accounted for. This is the relationship proposed by Terzaghi and Peck [18, article 54]. To apply the chart, one should read that reference.

Figure 38 shows charts for proportioning shallow footings on sand. This is a refinement over the older version of figure 37 by Peck [20, p. 309]. Again the design engineer needs to read the reference if performing a settlement analysis.

The SPT is widely used to evaluate liquefaction resistance of soils. Special procedures are required to perform SPT liquefaction evaluations [27]. Figure 39 shows the correlation between corrected blow count and

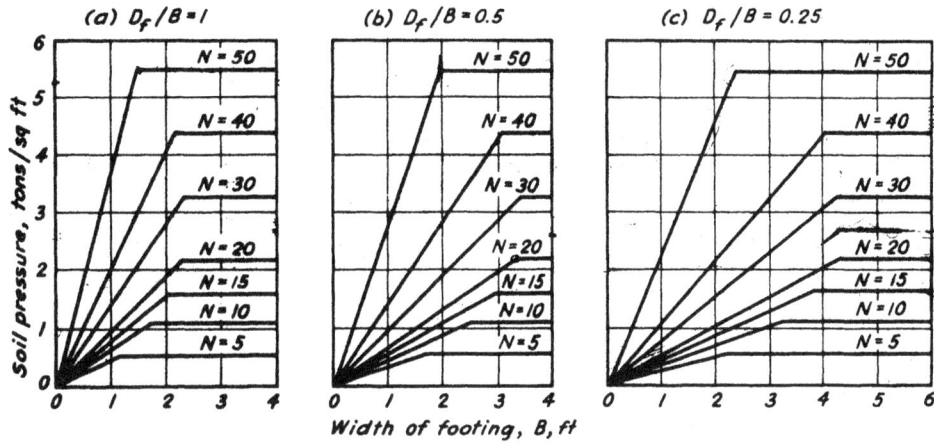

Figure 38.—Design chart for proportioning shallow footings on sand [20].

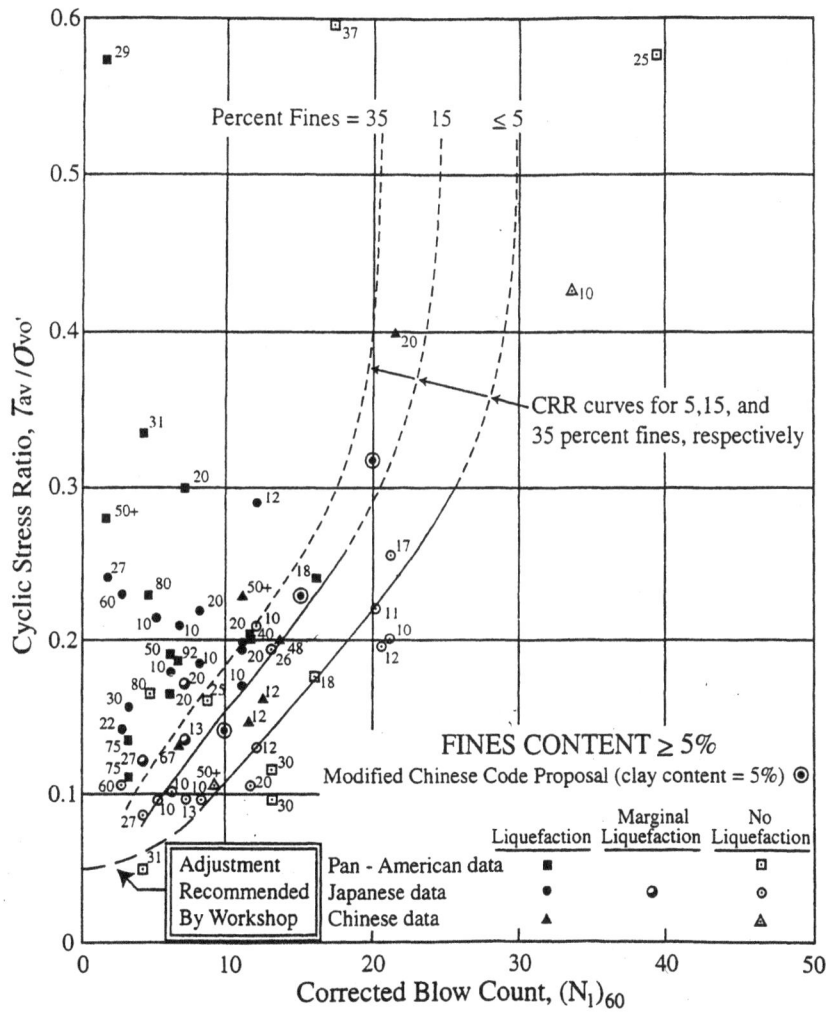

Figure 39.—Chart used to determine if sands are liquefiable.

cyclic stress ratio required to cause liquefaction [24].

The SPT sampler barrel is 1.37 inches in diameter and 24 inches in length, but the barrel is only driven 18 inches. The sample obtained is disturbed by driving and, depending on recovery, its mass is about 1 to 3 pounds. Physical properties and moisture contents are most often obtained on the sample. Moisture contents of clean sands are not reliable. Since the correlation between clay strength and SPT N is not very good, for clay samples, the moisture and Atterberg limits data are valuable to evaluate consistency. Very stiff clays have water contents close to their plastic limits, while very soft clays have moisture contents near their liquid limits.

3.4.3.1 Other Drive Sampling and Penetration Tests

Other drive samplers in use should not be confused with the SPT. ASTM standard practice D 3550 on thick wall, ring-lined, split barrel drive sampling of soils describes other methods for obtaining drive samples [28]. A popular version is known as the "California Barrel," which is a 3-inch outside diameter split barrel sometime equipped with brass rings. The brass rings hold specimens for laboratory testing, but driven samples are likely to be disturbed in many cases. All around the country, there are different combinations of hammers, drop heights, and sampler diameters. For example, certain State departments of transportation use a different penetration test than SPT and have developed local correlations of engineering parameters. These correlations may work well for the local geology but not be useful in other parts of the country.

One advantage of the larger barrel is the ability to recover large particles. Interest has been renewed in using "large penetration tests" (LPT) to obtain engineering properties of gravelly soils. However, no reliable method exists for correlating SPT to LPT.

3.4.3.2 Becker Penetration Test

The Becker drill consists of a double wall pipe six to seven inches in diameter that is driven with a double acting diesel hammer. This drill can penetrate and sample coarse alluvium. It was originally designed to prospect for gold and other mining/dredging applications. Recently it has been used to determine liquefaction resistance of soils containing gravels. SPT and CPT testing are not applicable or cannot be performed in gravels. For more information on the Becker Test for liquefaction evaluation, consult the Reclamations Seismic Design Standard.

3.4.4 Cone Penetration Testing

Cone penetration tests (CPT) are useful for determining relative density of granular soils, undrained strength of clays, and the stratigraphy of the underlining soils. CPTs, running about $10 to $15 per foot, are very cost effective relative to other drilling methods. CPT data result in very detailed stratigraphy as show on figure 40. To estimate engineering properties from CPTs, consult *Cone Penetration Testing in Geotechnical Practice* [29].

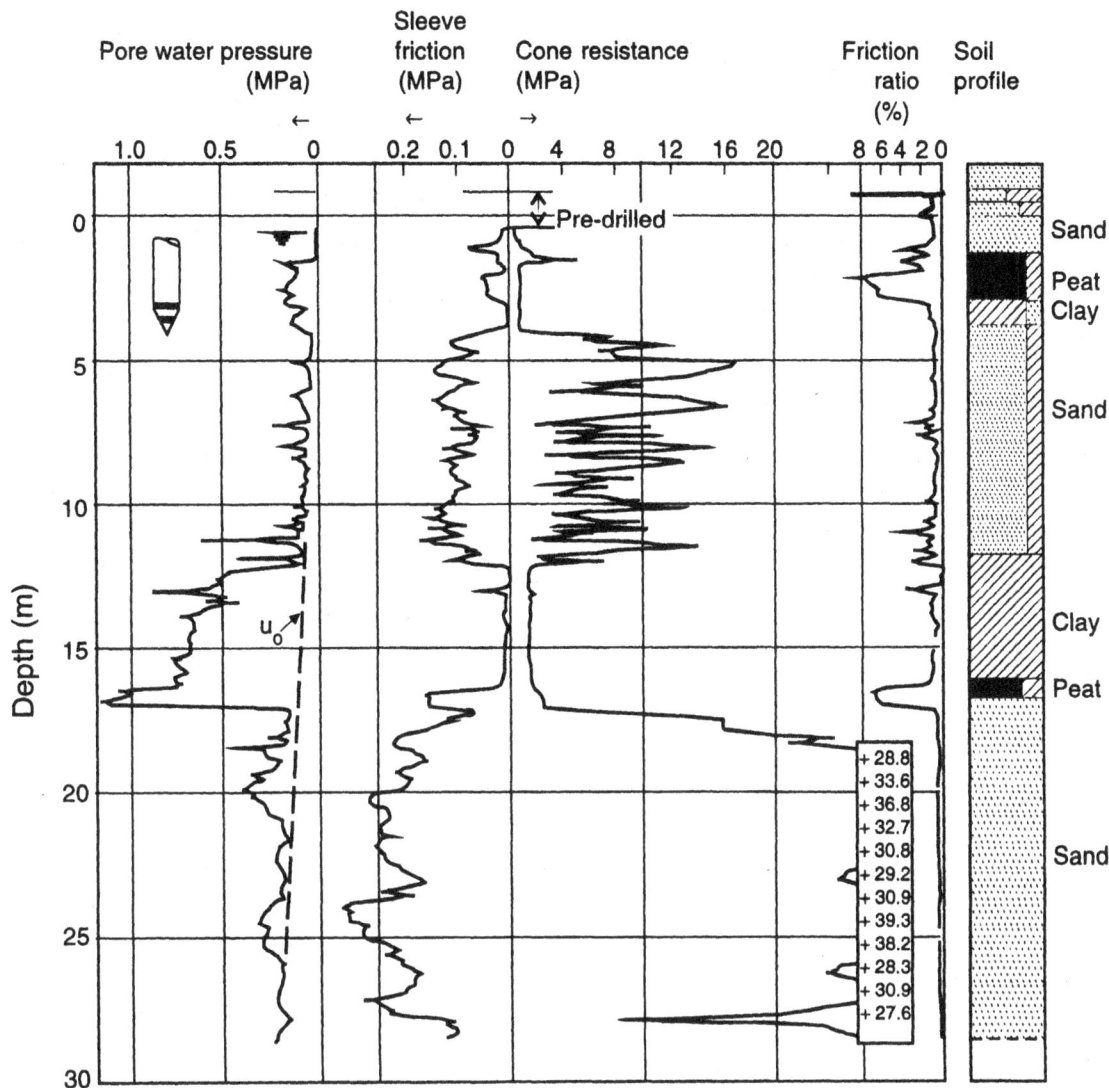

Figure 40.—Example CPTU results showing excellent profiling capability [29].

The CPT does not get a soil sample, but it can estimate soil type based the ratio of tip and sleeve resistance as shown on figure 41. The CPT estimates twelve soil behavior groups. The soil behavior groups are not exact matches to soil classification according to the unified classification system. Instead, ranges are given, such as "clayey silt to silty clay."

The CPT is excellent in distinguishing between clean sands and clays. It is a very popular tool for evaluating ground water flow conditions.

Clean sands have very high tip resistance and clays very low tip resistances. This is because at the constant penetration rate, sands are drained, and clays are undrained. Table 26 shows a relationship between permeability and soil behavior type. This correlation could be off by one order of magnitude.

For sands, both relative density and friction angle can be estimated. Figure 42 is one proposed relationship between CPT tip resistance data to effective overburden pressure

79

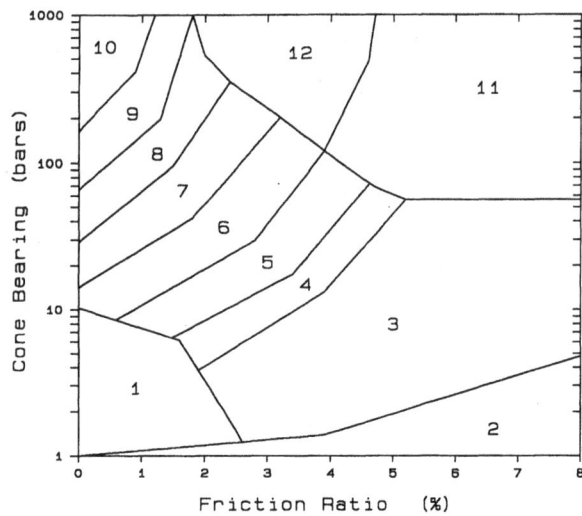

Zone	Qc/N	Soil Behaviour Type
1)	2	sensitive fine grained
2)	1	organic material
3)	1	clay
4)	1.5	silty clay to clay
5)	2	clayey silt to silty clay
6)	2.5	sandy silt to clayey silt
7)	3	silty sand to sandy silt
8)	4	sand to silty sand
9)	5	sand
10)	6	gravelly sand to sand
11)	1	very stiff fine grained (*)
12)	2	sand to clayey sand (*)

(*) overconsolidated or cemented

Figure 41.–Proposed soil behavior type classification system from CPTU data.

is one weakness in the method. For large investigations, an SPT boring can be located near a CPT sounding. Figure 44 shows one chart used for estimating liquefaction. Again, consult the seismic design standard [24] when performing liquefaction analysis.

Compressibilty of clay can be predicted by CPT according to table 27. The constrained modulus can be estimated based on soil type. This prediction is not very accurate, but may suffice for small structures. In some cases, the prediction can be supplemented by information from soil samples.

CPT is an excellent predictor of undrained strength, S_u, of clays, because the clay fails rapidly beneath the cone. Comparisons between field vane shear, and laboratory unconsolidated undrained tests to CPT derived undrained strength prediction have been excellent. For small structures, the prediction of S_u from CPT alone may suffice. Consult reference [29] for methods to predict undrained strength. The typical method is to use the total stress approach.

and shear strength. Figure 43 shows one chart that correlates CPT data to relative density and vertical effective stress. These charts are developed for clean quartz sands. Sands have differing compressibilities, depending on their mineralogy and the fines added. For example, 5 to 10 percent mica makes clean quartz sand much more compressible.

The CPT is frequently used for liquefaction resistance evaluations, much like the SPT. The CPT is considered to be more reliable than the SPT, because the test is less subject to error. However, the method requires estimating fines content, and if no soil samples are available, this

3.4.5 Drilling—Power and Bucket Augers

Disturbed soil samples procured by high speed solid stem power augers [2, USBR 7010] and bucket augers [1] provide a rapid, economical means of confirming soil conditions along line structures and in borrow areas. These exploration methods can be used to fill in the gaps between test pits and drill holes. Even if soils are mixed, changes in strata can be detected, and moisture content data can be taken. In general, soils can be grouped in 5-foot depth increments, classified, and samples bagged for testing if required.

Table 26.—Estimation of soil permeability, k, from CPT soil behavior charts [29]

Zone	Soil behavior type (SBT)	Range of soil permeability k (m/s)
I	Sensitive fine grained	3×10^{-9} to 3×10^{-8}
2	Organic soils	1×10^{-8} to 1×10^{-6}
3	Clay	1×10^{-10} to 1×10^{-9}
4	Silty clay to clay	1×10^{-9} to 1×10^{-8}
5	Clayey silt to silty clay	1×10^{-8} to 1×10^{-7}
6	Sandy silt to clayey silt	1×10^{-7} to 1×10^{-6}
7	Silty sand to sandy silt	1×10^{-5} to 1×10^{-6}
8	Sand to silty sand	1×10^{-5} to 1×10^{-4}
9	Sand	1×10^{-4} to 1×10^{-3}
10	Gravelly sand to sand	1×10^{-3} to 1
11	*Very stiff fine-grained soil	1×10^{-9} to 1×10^{-7}
12	*Very stiff sand to clayey sand	1×10^{-8} to 1×10^{-6}

*Overconsolidated or cemented

Table 27.—Estimation of constrained modulus, M, for clays [29]

$M = 1/m_v = \alpha_m q_c$		
$q_c < 0.7$ MPa	$3 < \alpha_m < 8$	Clay of low plasticity (CL)
$0.7 < q_c < 2.0$ MPa	$2 < \alpha_m < 5$	
$q_c > 2.0$ MPa	$1 < \alpha_m < 2.5$	
$q_c > 2$ MPa	$3 < \alpha_m < 6$	Silts of low plasticity (ML)
$q_c < 2$ MPa	$1 < \alpha_m < 3$	
$q_c < 2$ MPa	$2 < \alpha_m < 6$	Highly plastic silts and clays (ME, CH)
$q_c < 1.2$ MPa	$2 < \alpha_m < 8$	Organic silts (OL)
$q_c < 0.7$ MPa	$1.5 < \alpha_m < 4$	Peat and organic clay (P_t, OH)
$50 < w < 100$	$1 < \alpha_m < 1.5$	
$100 < w < 200$	$0.4 < \alpha_m < 1$	
$w > 200$		

w = water content

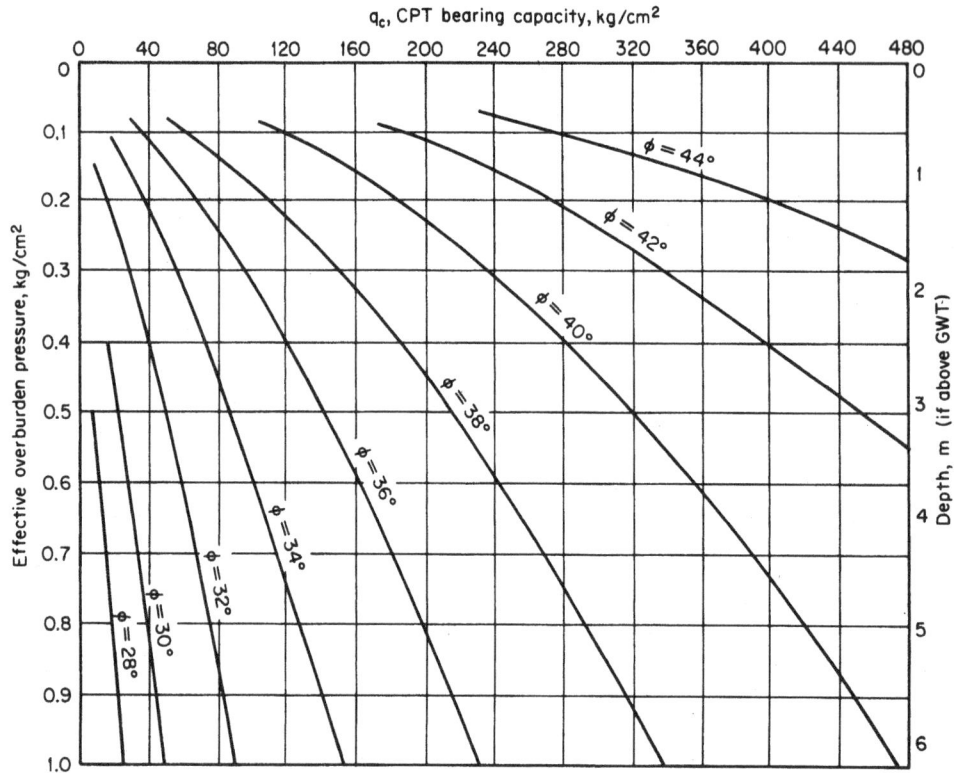

Figure 42.—Correlation between effective overburden pressure, q_c and ϕ [14].

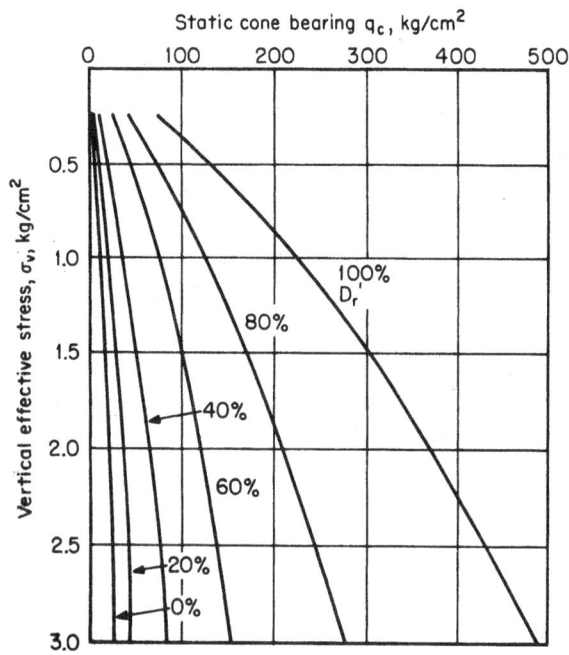

Figure 43.—Static cone resistance [14].

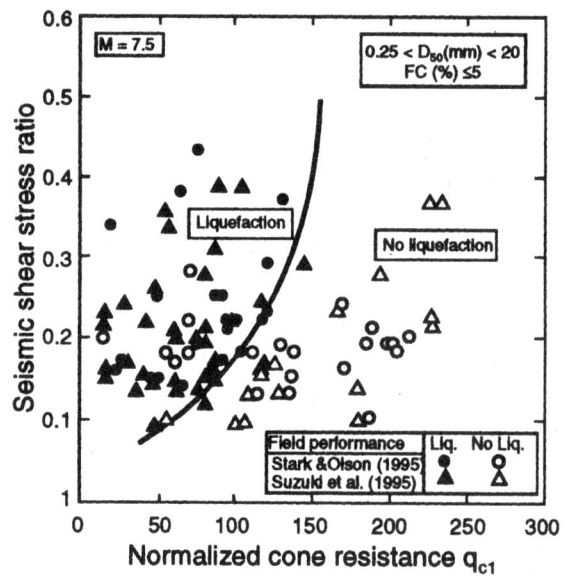

Figure 44.—Recent field evidence of cyclic liquefaction with CPT resistance curve for clean sand [29].

3.4.6 Rotary Drilling—Hollow-Stem Auger

A predominant rotary drilling method in use today is the hollow-stem auger (HSA). USBR 7105 [2], provides information on procuring undisturbed samples for the laboratory using the HSA. Recently, a new ASTM Standard D 6151 [30] has been issued, which provides much more detail on the operation of these systems for geotechnical exploration. Reclamation favors the hollow-stem auger because its use avoids hydraulic fracturing in dams. Also, time savings are achieved, because drilling fluids are not required. For undisturbed sampling, the HSA can be equipped with acrylic liners in 3.25- and 5.25-inch diameter sizes (fig. 45). The HSA has virtually replaced Denison and Pitcher barrel sampling in Reclamation.

Two types of HSA systems can be used, rod type and wireline type. The wireline systems are much faster to operate under certain ground conditions. In dry soils and in cohesive soils, the wireline systems work well. However, in sands below the water table, wireline systems have problems latching, and rod type systems may be required. Rod type systems are often used for large diameter coring to prevent core spinning, but they are slower to operate.

HSA systems come equipped with inner tube core barrels for simultaneous drilling and sampling. Core diameters range from 3 to 6 inches. The core barrels can be split barrels with cutting shoes, or the inner barrel can be equipped with a liner for undisturbed sampling. The majority of sampling for small structures can be performed with 3- to 4-inch split barrels. Considering depth limitations of test pits, HSA drilling provides an economical alternative and allows for adequate exploration depths of small

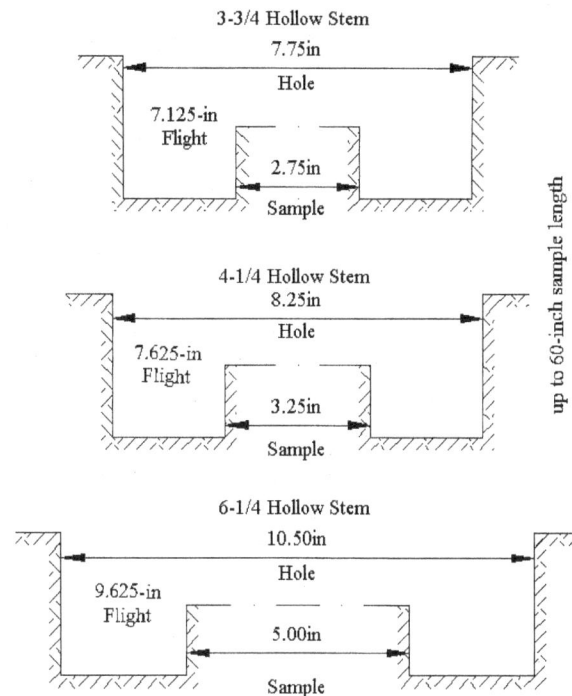

Figure 45.—The dimensions of undisturbed samples that can be obtained from various hollow stem auger systems.

structures. Since no drilling fluid is used, HSA is advantageous in sampling of collapsible soils.

Cores of clays can be examined for structure, and simple tests such as pocket penetrometers or Torvane [2, USBR 5770] tests can be performed for quick indications of bearing capacity. Some clay samples should be tested for moisture and consistency (Atterberg Limits) [2] because there are methods to estimate the compression index of clays based on in-place moisture and consistency of clay.

Cores of sands and gravels will likely be disturbed, and basket catching devices are sometimes needed for recovery. HSA systems are often combined with penetration resistance tests for foundation investigations. The HSA system is known to have considerable problems with disturbance when used with PR tests in sands below the water table. Consult the SPT driller guide [26] for more information on how to avoid disturbance in saturated sands. In free draining sands and gravels, determination of the moisture content of the cores is unnecessary, because they are often unreliable due to drainage during sampling.

PR tests are often performed at set intervals or changes in materials. Between these intervals, the HSA coring system can be used to core over and in between PR tests.

Reclamation uses the hollow-stem auger system to obtain relatively undisturbed soil cores for laboratory testing. The inner split barrel can be equipped with liners and special cutting shoes. The procedure is described in USBR.7105 and ASTM D 6151[2, 30].

3.4.7 Rotary Drilling—Samplers

Many other rotary drilling methods, such as fluid or air rotary, and casing advancers, are available for exploration, but will not be addressed in this manual. For detailed information on a wide array of drilling methods consult the *Earth Manual* [1, 2].

For very soft clay soils, the HSA sampler may cause significant disturbance, and the use of thin wall tubes (ASTM D 1587) should be considered. For difficult-to-recover silts and sands, piston samplers, which use thin wall tubes, are sometimes required. Thin wall tubes come in sizes ranging from 3 to 5 inches. Samples taken in thin wall tubes should be tested promptly, as soil reacts adversely to contact with the metal tube.

3.4.8 Rotary Drilling—Diamond Rock Coring

Diamond rock coring is rarely used for investigations of small structures. In many cases, the geologist can describe the rock types to be encountered based on experience with the site. If rock coring is required, consult the *Earth Manual* [1, 2]and *Engineering Geology Field Manual* [3] for information on drilling and coring. Reclamation has recently upgraded the ASTM diamond drilling standard D 2113 [31], and the core barrel tables are more complete than those in the *Earth Manual* [2].

Rock generally makes a sufficient foundation for light structures. Rock investigations are required for tunneling operations and may be required in areas of slope stability concerns. Rock can range from extremely soft to extremely hard. Table 28 shows a method of classifying intact rock.

A key rock parameter is its uniaxial compressive strength. Figure 46 shows a correlation between uniaxial compressive strength and deformation modulus. For tunneling and microtunneling applications, the key information needed is the compressive strength and rock hardness for the contractor to evaluate equipment wear. If rock compressive strength is required, consider using less expensive, indirect tests, such as a the point load test. The indirect tests can be correlated to lab compressive strengths, and the amount of expensive lab testing can be reduced to save investigation costs.

To estimate engineering properties of rock, engineers use the rock quality designation (RQD), and have developed rock mass rating (RMR) systems for applications such as

Table 28.—Intact rock classification based on hardness and weathering [22]

Class	Hardness [1]	Diagnostic features ( ) = Weathering effects	Weathering grade [2]	Symbol	Strength, [3] t/ft^2
I	Extremely hard (or strong)	Rings under hammer impact; many blows required to break specimen. (No visible signs of decomposition or discoloration.)	Fresh	F	> 2500
II	Very hard to hard (or very strong)	Hand-held specimen breaks with hammer under more than one blow. (Slight discoloration inward from open fractures, otherwise similar to F.)	Slightly weathered	WS	2500-1000
III	Moderate (or medium strong)	Cannot be scraped or peeled with knife. Hand-held specimen can be broken with single moderate hammer blow. (Discoloration throughout; weaker minerals, such as feldspar, decomposed. Texture preserved.)	Moderately weathered	WM	500-250
IV	Soft (or weak)	Can just be scraped or peeled with knife. Indentations 1 to 3 mm show in specimen with moderate blow with pick end; lower strength specimens can be broken by hand with effort. (Most minerals somewhat decomposed; texture becoming indistinct but fabric preserved.)	Highly weathered	WH	250-50
V	Very soft (or very weak)	Material crumbles under moderate blow with pick and can be peeled with knife, but is hard to hand-trim for test specimen. (Minerals decompose to soil but fabric and structure preserved; i.e., saprolite.)	Completely weathered	WC	50-10
	Extremely soft or weak	Advanced state of decomposition	Residual	RS	<10

[1] Hardness depends on rock type, as well as weathering grade.
[2] Weathering grade applies primarily to crystalline rocks.
[3] Relationships to be considered only as a general guide, from U_c test.

underground tunneling, mining excavation, and blasting slope stability. Table 29 is an example of a rating system for rock rippability. For more information, consult your geologist on the need for RMR.

85

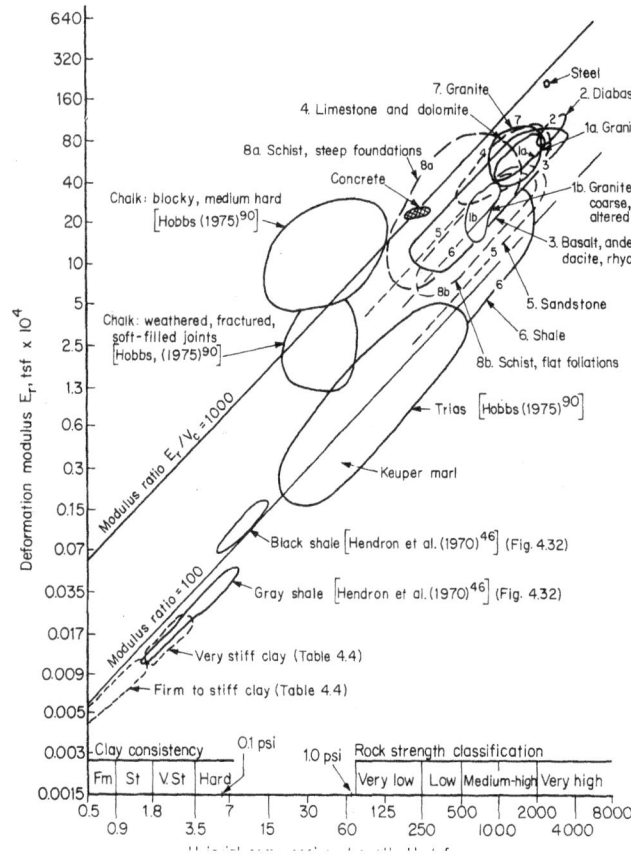

Figure 46.—Relationships between uniaxial compressive strength and deformation modulus for various rock types and clays [22].

Table 29.—Rippability classification chart [32]

Parameters	Class 1	Class 2	Class 3	Class 4	Class 5
Uniaxial tensile strength (MPa)	<2	2-6	6-10	10-15	>15
Rating	0-3	3-7	7-11	11-14	14-17
Weathering	Complete	Highly	Moderate	Slight	None
Rating	0-2	2-6	6-10	10-14	14-18
Sound velocity (m/s)	400-1100	1100-1600	1600-1900	1900-2500	>2500
Rating	0-6	6-10	10-14	14-18	18-25
Abrasiveness	Very low	Low	Moderate	High	Extreme
Rating	0-5	5-9	9-13	13-18	18-22
Discontinuity spacing (m)	<0.06	0.06-0.3	0.3-1	1-2	>2
Rating	0-7	7-15	15-22	22-28	28-33
Total rating	<30	30-50	50-70	70-90	>90
Ripping assessment	Easy	Moderate	Difficult	Marginal	Blast
Recommended dozer	Light duty	Medium duty	Heavy duty	Very heavy duty	

Chapter 4

Problem Soils

4.1 Silts and Low Plasticity Soils

Silts and low plasticity clays are soils with moderate characteristics. They generally are not considered a problem in a recompacted condition, except they may need to be checked for settlement and shear resistance by laboratory tests. Natural deposits, however, are frequently a problem, because they are sometimes very loose or of low density, such as windblown

(Loess) deposits of the Midwest (Kansas, Nebraska) and Washington State, and slopewash/mudflow deposits of the San Joaquin Valley. Table 30 and figure 47 show the locations of windblown soils and other soil types in the United Sates [17].

Silt and low plasticity clays can be very stable under dry conditions, where small amounts of clay binder hold them together. As these soils become saturated, however, the binding effect

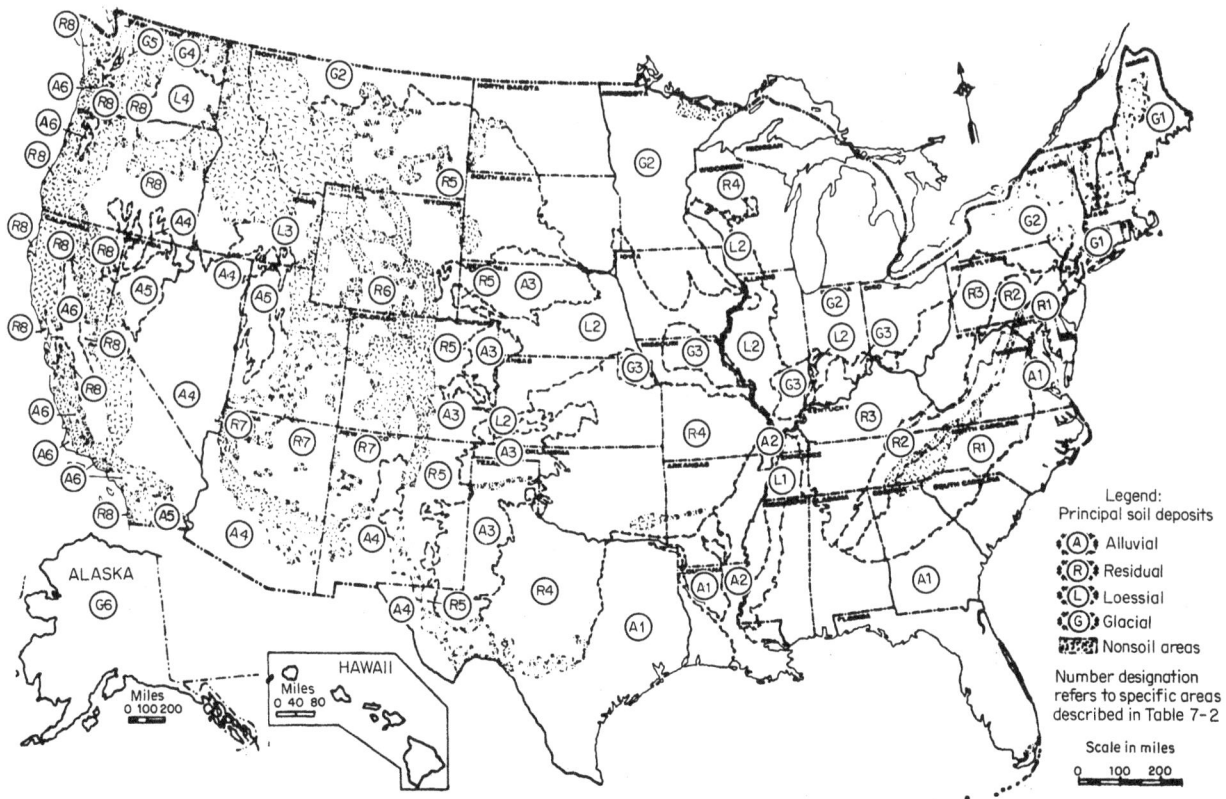

Figure 47.—Distribution of soils in the United States classed by origin [17].

Table 30.—Distribution of principal soil deposits in the United States.

Origin of principal soil deposits	Symbol for area in fig. 47	Physiographic province	Physiographic features	Characteristic soil deposits
Alluvial	A1	Coastal plain	Terraced or belted coastal plain with submerged border on Atlantic. Marine Plain with sinks, swamps, and sand hills in Florida.	Marine and continental alluvium thickening seaward. Organic soils on coast. Broad clay belts west of Mississippi. Calcareous sediments on soft and cavitated limestone in Florida.
Do	A2	Mississippi alluvial plain	River floodPlain and delta.	Recent alluvium, fine grained and organic in low areas, overlying clays of coastal plain.
Do	A3	High Plains section of Great Plains province	Broad intervalleY remnants ou smooth fluvial plains.	Outwash. mantle of silt, sand, sit clay, lesser ravels underlain b soft shale, sandstone, and marls.
Do	A4	Basin and range province	Isolated ranges of dissected block mountains separated by desert plains.	Desert plains farmed principally of alluvial fans of coarse-grained soils merging to playa lake deposits. Numerous nonsoil areas.
Do	A5	Major lakes of basin and range Province	Intermontane Pleistocene lakes in Utah and Nevada, Salton Basin in California.	Lacustrine silts and clays with beach sands on periphery. Widespread sand areas in Salton basin.
Do	A6	Valleys and basins of Pacific border province	Intermontane lowlands, Central Valley, Los Angeles Basin, Willamette Valley.	Valley fills of various gradations, fine grained and sometimes organic in lowest areas near drainage system.
Residual	Rl	Piedmont Province	Dissected peneplain with moderate relief. Ridges on stronger rocks.	Soils weathered in place from metamorphic and intrusive rocks (except red shale and sandstone in New Jersey). Generally more clayey at surface.
Do	R2	Valley and ridge province	Folded strong and weak strata forming successive ridges and valleYs,	Soils in valleys weathered from shale, sandstone and limestone. Soil thin or absent on ridges.
Do	R3	Interior low plateaus and Appalachian plateaus	Mature, dissected plateaus of moderate relief.	Soils weathered in place from shale, sandstone, and limestone.

88

Table 30.—Distribution of principal soil deposits in the United States.

Origin of principal soil deposits	Symbol for area in fig. 47	Physiographic province	Physiographic features	Characteristic soil deposits
Do	R4	Ozark Plateau, Ouachita province, Portions of Great Plains and central lowland Wisconsin driftless section	Plateaus and Plains of moderate relief, folded strong and weak strata in Arkansas.	Soils weathered in place from sandstone and limestone predominantly, and shales secondarily. Numerous nonsoil areas in Arkansas.
Do	R5	Northern and western sections of Great Plains province	Old plateau, terrace lands and Rocky Mountain piedmont.	Soils weathered in place from shale, sandstone and limestone including areas of clay-shales in Montana South Dakota, Colorado.
Do	R6	Wyoming basin	Elevated Plains.	Soils weathered in place from shale sandstone, and limestone.
Do	R7	Colorado Plateaus	Dissected Plateau of strong relief	Soils weathered in place from sandstone primarily, shale and limestone secondarily.
Do	R8	Columbia Plateaus and Pacific border province	High Plateaus and piedmont.	Soils weathered from extrusive rocks in Columbia plateaus and from shale and sandstone on Pacific border. Includes area of volcanic ash and pumice in central Oregon.
Loessial	L1	Portion of coastal plain	Steep bluffs on west limit with incised drainage,	30 to 100 ft of loessial silt and sand overlying coastal plain alluvium. Loess cover thins eastward.
Do	L2	Southwest section of central lowland; Portions of Great Plains	Broad intervalleY remnants of smooth plains.	Loessial silty clay, silt, silty fine sand with clayey binder in western areas, calcareous binder in eastern areas.
Do	L3	Snake River Plain of Columbia plateaus	Young lava plateau.	Relatively thin cover of loessial silty fine sand overtYing fresh lava flows,
Do	L4	Walla Walla Plateau of Columbia plateaus	Rolling plateau with young incised valleys.	Loessial silt as thick as 75 ft overlying basalt. Incised valleys floored with coarse-grained alluvium.
Glacial	G1	New England Province	Low PenePlain maturely eroded and glaciated,	Generally glacial till overlying metamorphic and intrusive rocks, frequent and irregular outcroPs. Coarse, stratified drift in upper drainage systems. Varved silt and clay deposits at Portland, Boston, New York, Connecticut River Valley, Hackensaok area.

Table 30.—Distribution of principal soil deposits in the United States.

Origin of principal soil deposits	Symbol for area in fig. 47	Physiographic province	Physiographic features	Characteristic soil deposits
Do	G2	Northern section of Appalachian plateau, Northern section of Central lowland	Mature glaciated plateau in northeast, young till plains in western areas.	Generally glacial till overlying sedimentary rocks. Coarse stratified drift in drainage system. Numerous swamps and marshes in north central section. Varved silt and clay deposits at Cleveland, Toledo, Detroit, Chicago, northwestern Minnesota,
Do	G3	Areas in southern central lowland	Dissected old till plains.	Old glacial drift, sorted and unsorted, deeply weathered overlying sedimentary rocks.

deteriorates, and the soils become unstable and frequently subside appreciably; this is shown on figure 48. Spoil banks and uncompacted dump fills will most always fall into the category of collapsible soils. The figure shows a canal built in Washington State with subsequent collapse of Loess. It is important that the looseness of these soils be recognized. One method of treatment is to pond the foundation area prior to construction as shown on figure 49. Another economical method feasible for small structures is dynamic compaction.

To evaluate collapse, the density of the soil must be measured. Density tests can be taken in test pits, on block samples, and from large diameter hollow-stem core samples. Two criteria can be checked as shown on figure 50. The criteria are based on the in situ dry unit weight, laboratory compaction, and liquid limit of the material.

Soils of different plasticity or water-holding capacity will collapse at different densities. The liquid limit [2, USBR 5350] is a moisture content, determined by standard laboratory tests, which represents the weakest plastic condition of the soil or the "approaching a liquid" condition. When the soil has a low density, such that its void space is sufficiently large to hold the liquid limit moisture content, saturation can easily cause a liquid-limit consistency and the soil is able to collapse. When a soil has a void space too small to hold the liquid-limit moisture, it cannot reach this consistency even when it becomes saturated. It will not collapse but will retain a plastic condition and only settle as a normal result of loading.

When the natural density is evaluated at frequent locations and depths and compared to the limiting density based on the liquid-limit moisture content or degree of compaction, the

Figure 48.—Cracking and settling of canal bank in dry, low density silt [1].

Figure 49.—Ponding dry foundation of Trenton Dam in Nebraska [1].

criterion discussed above becomes a useful aid in showing when in-place densities are either adequate or lower than limiting density and trend toward probable near-surface subsidence.

It is also of value to correlate the density relations to such other characteristics as degree of saturation, depth to ground water, ponding tests at representative locations, and laboratory tests on representative samples to evaluate loading effects. These methods were part of the San Luis Canal investigations, California, where subsidence is a serious problem.

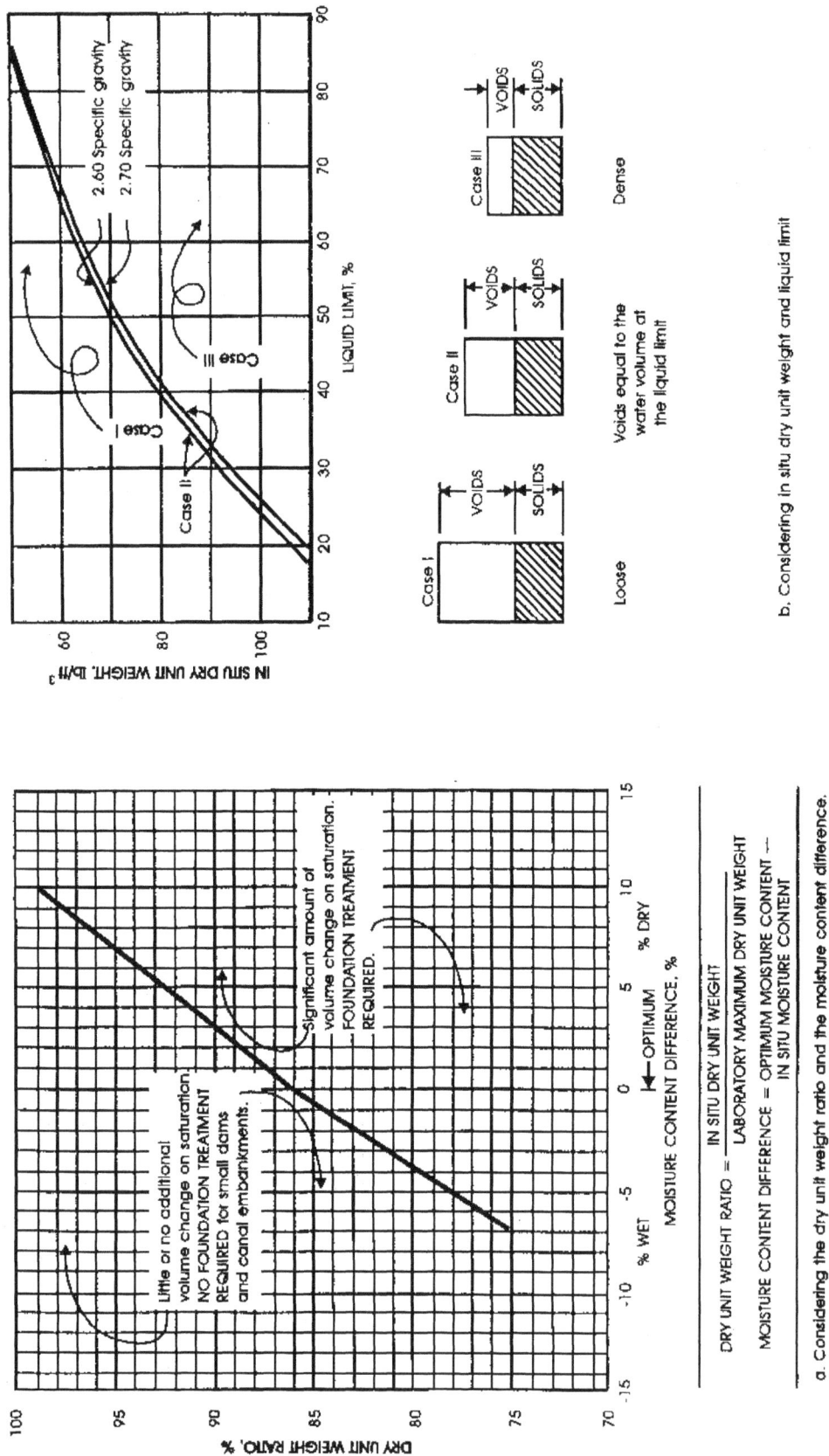

Figure 50.—Criterion for treatment of relatively dry, fine grained foundations [1].

4.2 Clays

4.2.1 Firm, Consolidated Clays

As a foundation material, firm, consolidated clays are generally desirable. They are frequently called formation material and in geologic terms, may even be called bedrock. Examples of such materials are the Denver Formation in Denver and Carlyle shale, Pierre shale, and Colorado shale found in the Midwest. These clays have been consolidated by large, formerly overlying soil or glacial ice pressures and, as a result, are preconsolidated to pressures greater than those placed on them by new structures. Also, such materials located below the ground surface are desirable for supporting piles and caissons. As mentioned previously, they may be subject to deterioration upon disturbance. However, as long as they are not disturbed, the preconsolidation pressures have given them strengths that are capable of sustaining appreciable loads. These soils may swell when the excavation process removes the overburden loads. Also, if they contain expansive clay minerals, the influx of water may make them objectionable. This will be discussed in section 4.2.3. Nevertheless, for moderate-sized structures, the supporting capacity of such clays is usually not a problem. The location and general firmness of such clays can be determined by penetration resistance tests, undrained strength tests, and simple thumbnail penetration tests (refer to sec. 3.2.5).

Stiff, fissured, overconsolidated clays and clay shales containing weak clay seams have been problematic in cut slope stability. The shale formations in the western U.S. associated with weak bentonitic seams are the Pierre and Mancos shales. Very stiff, overconsolidated clays can also be a problem. If displacement

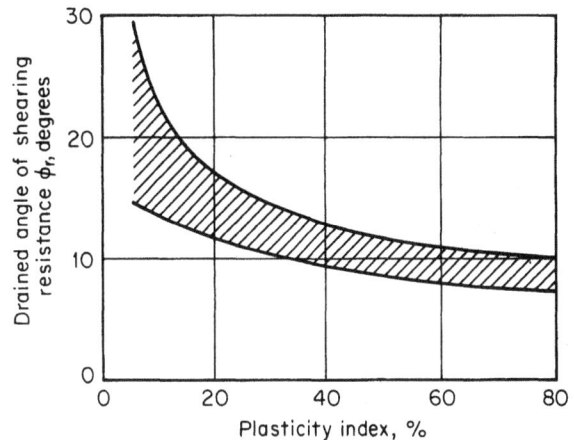

Figure 51.—Approximate relationship between the drained angle of residual shearing resistance and plasticity index for rock gouge material [14].

has occurred on shear zones or seams, a resulting low residual friction angle of 8 to 10 degrees can result. Sedimentary shales with unfavorable bedding attitudes in cut slopes should be scrutinized for weak seams. Reworked clay gouge in rock faults should also be surveyed. Repeated direct shear tests can be performed to measure the residual strength. Fortunately, the possible residual strength can be estimated from Atterberg limit data alone. Figure 51 shows one relationship for the residual strength of clay gouge.

4.2.2 Compressible Clays

Compressible clays are critical materials regarding foundation settlement and bearing capacity. These are the materials that require soil mechanics analyses and decisions of whether settlement can be tolerated or whether the foundation should be improved by excavating and placing the foundation at greater depth, or by the use of piles or caissons, or the use of a compacted earth pad. These clays are considered "Normally Consolidated" and have not been preconsolidated by heavy loads in the past. Instead, they are geologically recent deposits and, commonly, new water-deposited sediments. An extreme example is the lake sediments near the Great Salt Lake, Utah. Arthur V. Watkins Dam near Ogden, Utah, was placed on this soft clay, and the rather large settlements of several feet that are taking place were accounted for in the design. The dam embankment was built using the stage construction method. However, rigid structures such as the pumping plants of this project are bypassing this clay with deep piles.

Compressible clays are not necessarily as compressible as the above example but may vary up to what are called the preconsolidated or firm consolidated clays. Compressible clays are not excluded as satisfactory foundations but are the clays that usually require tests and soil mechanics interpretation for the evaluation of their supporting capacity. They may be critical in both settlement and the problem of punching into the ground, accompanied by lateral bulging. The field penetration resistance test is a good, rapid, and preliminary method of exploring the general quality of such a foundation. Although this test shows the degree of firmness, it does not indicate settlement probability. It does, however, indicate shearing characteristics of the soil in its in-place condition, but changes in moisture may appreciably affect the strength, and this effect is not indicated by an in-place test. In the case of very soft, saturated clays, the vane test [2, USBR 7115] or cone penetration test are more precise methods of evaluating the shearing strength of the material in place (refer to sections 3.2.3.5 and 3.2.3.7). The most acceptable evaluation of the settlement and bearing capacity characteristics is obtained from undisturbed samples and laboratory tests. When detailed tests are not practicable, in-place density and laboratory index property tests are very valuable to supplement simplified field tests such as penetration resistance.

4.2.3 Expansive Clays

The expansion of some clay and clay shales when additional water is made available to them is a characteristic that is related to the type of minerals composing the clays. This can be a problem with either compressible or consolidated clays. It is, of course, far more critical when the clays are initially dense, as in consolidated clays, because they then have much more possibility of expansive volume change.

These clays have mineral constituents with an affinity for pulling water molecules into their structure. The water effect on these clays causes them to shrink and crack when they are dried, and swell with sometimes-appreciable force when they become wetted. Such clays usually fall in the CH (fat clay) classification group. Devising methods of identifying these clays has been an aid to many organizations, and Reclamation's guideline in table 31 has received much recognition.

Use of this table requires data from index property tests [2, USBR 5330, USBR 5350, USBR 5360, USBR 5365] on the basis of the following principles:

- A high plasticity index means that the soil can have a large change in moisture and still be in a plastic condition. Since moisture is the principal cause of volume swell, it is logical that a high plasticity index is a contributing indicator of expansion.

- A low shrinkage limit value means that it is possible that the soil can shrink to small volumes by drying. Therefore, this characteristic would be a

Table 31.—Relation of soil index properties to expansion potential of high-plasticity clay soils. Data for making estimate of probable volume change for expansive materials

Data from index tests [1]			Probable expansion [2], percent total volume change, dry to saturated condition	Degree of expansion
Colloid content, percent minus 0.001 mm	Plasticity index, PI, %	Shrinkage limit, SL, %		
>28	>35	<11	>30	Very high
20 to 31	25 to 41	7 to 12	20 to 30	High
13 to 23	15 to 28	10 to 16	10 to 20	Medium
<15	<18	>15	<10	Low

[1] All three index tests should be considered in estimating expansive properties.
[2] Based on a vertical loading of 7 kPa (1.0 lbf/in^2)

contributing indicator of possible shrinking volume change.

- A soil with a high content of colloidal clay as the most active ingredient means that a large amount of material in the soil has the possibility of causing expansion.

The table is based on tests on many samples and serves as a guide to estimating the percentage and degree of expansion. In the use of this table, all three of the above-mentioned properties should be considered together to arrive at the expansion estimate.

Data in table 31 will identify high-volume-change clays during investigations. However during construction, the in-place moisture density condition is most important. On San Luis Drain, if the natural moisture of a soil were sufficiently high, it would not expand thereafter if the moisture were retained. Figure 52 shows the effect of increasing water content when clays expand during laboratory testing. From these data, a boundary of minimum water content for a required soil liquid limit was

determined to stabilize clays for hydraulic structure foundations.

If the natural clays dry during construction, they are moistened by sprinkling for 30 days before embankments are built on them. Expansive clays can be controlled from a structure foundation standpoint by (1) placing the structure on caissons to increase foundation loadings and to anchor it in nonchanging material, (2) controlling subsoil moisture changes, and (3) mixing lime (sec. 4.18.2) with the clay to change the chemical structure and form a stronger product.

Expansive clays are particularly objectionable for hydraulic structures, due to the cyclical wet and drying that occurs.

Figure 53 shows an example of soil expanding below a canal lining. Rigid linings are not successful for canals to be built in expansive soils. If expansive soils are anticipated, use of a flexible membrane liner should be considered.

Heave of the subgrade can also be caused by an excessive amount of sodium sulfate salt in the

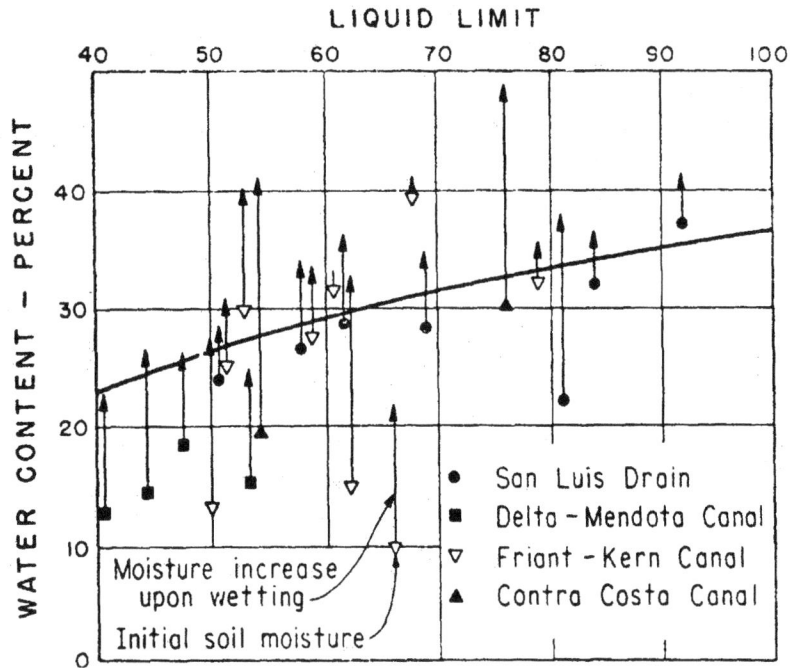

(a) Increase in water content when clays expanded during laboratory tests.

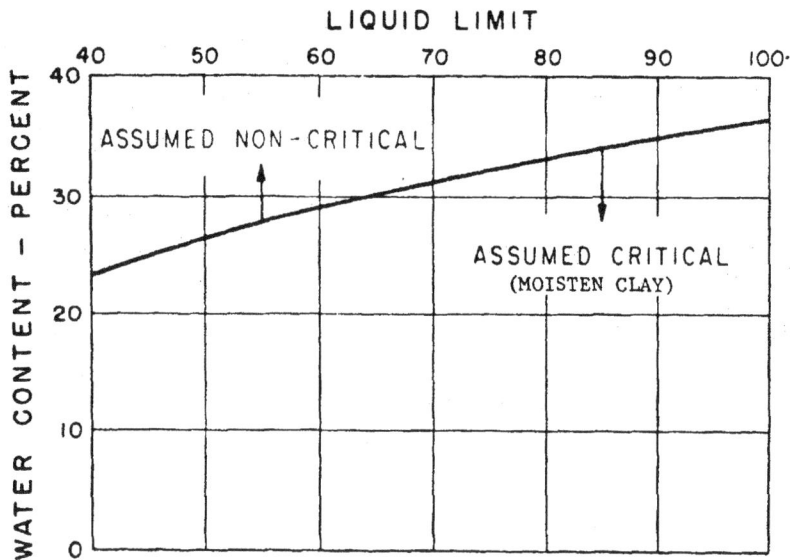

(b) Minimum water content required for soil liquid limit to stabilize
clays for hydraulic structure foundations.

Figure 52.—Development of moisture control criterion for reducing
expansive potential of foundation clays during construction.

pore water that crystallizes due to temperature changes.

To further assess the expansion potential of soil, physical property tests can be performed. Laboratory expansion and uplift tests may be required [2, USBR 5705, USBR 5715].

4.3 Sands

Granular and gravelly soils are generally more desirable for foundations. About the only major undesirable feature of these soils is their occasional occurrence at low density. Reclamation uses the relative density test [2, USBR 5525, USBR 5530] to find out how dense these soils are. This test shows the natural density in terms of the minimum and maximum density that can be obtained by a standard method in the laboratory. Normally, the soil density is acceptable if it is above 70 percent relative density.

4.4 Liquefaction

If sandy soils are loose and saturated, they may liquefy under earthquake shocks. Liquefaction is associated with sands, silty sands, and silts. Clayey soils, in general, are not liquefiable. Liquefaction damages include settlement after sand boiling, lateral spreading, and slope failures. In some cases, liquefaction and subsequent deformation and settlement of the structure are acceptable risks, as opposed to expensive foundation treatments. In other cases, liquefaction must be evaluated and liquefiable soils either treated or removed and replaced. Information regarding foundation improvement methods can be found in the *Earth Manual* [1, 2]. For light structures, treatment may consist of simple excavation and replacement with an impervious soil.

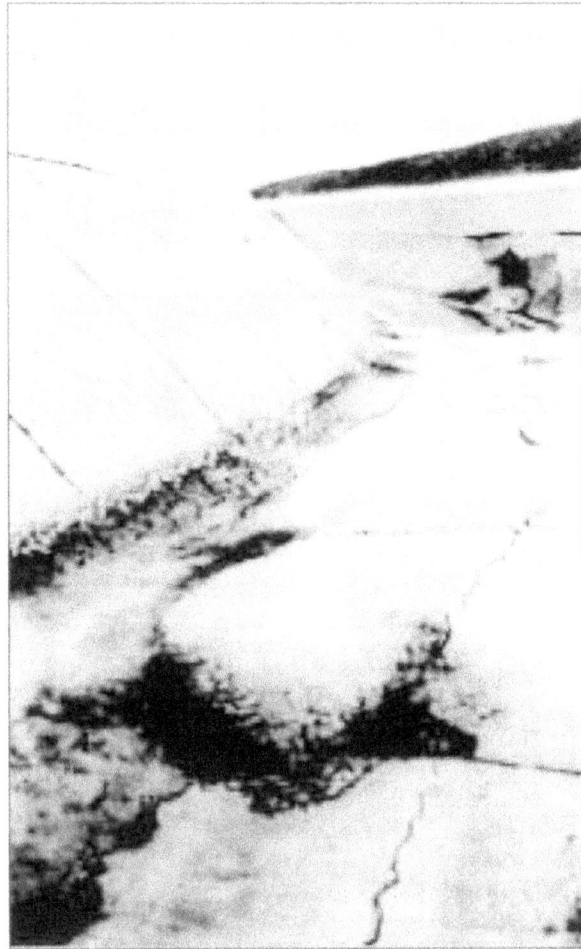

Figure 53.—Soil expanding below a canal lining.

Liquefaction can be evaluated through penetration tests (sections 3.4.3 and 3.4.4) or in-place relative density determination. Table 32 shows types of sand deposits and their estimated liquefaction susceptibility. Virtually all very young sand deposits are susceptible to liquefaction, while older deposits are more resistant.

4.5 Gravels

Gravelly soils are considered particularly desirable as a construction material; they are the select materials used for roadways. Gravel has advantages of low compressibility or settlement

Table 32.—Liquefaction susceptibility of soil deposits [22]

Type of deposit (1)	General distribution of cohesionless sediments in deposits (2)	Likelihood that cohesionless sediments, when saturated, would be susceptible to liquefaction (by age of deposit)			
		< 500 yr (3)	Holocene (4)	Pleistocene (5)	Prepleistocene (6)
(a) Continental deposits					
River channel	Locally variable	Very high	High	Low	Very low
Flood plain	Locally variable	High	Moderate	Low	Very low
Alluvial fan and plain	Widespread	Moderate	Low	Low	Very low
Marine terraces and plains	Widespread	-	Low	Very low	Very low
Delta and fan-delta	Widespread	High	Moderate	Low	Very low
Lacustrine and playa	Variable	High	Moderate	Low	Very low
Colluvium	Variable	High	Moderate	Low	Very low
Talus	Widespread	Low	Low	Very low	Very low
Dunes	Widespread	High	Moderate	Low	Very low
Loess	Variable	High	High	High	Unknown
Glacial till	Variable	Low	Low	Very low	Very low
Tuff	Rare	Low	Low	Very low	Very low
Tephra	Widespread	High	High	?	?
Residual soils	Rare	Low	Low	Very low	Very low
Sabka	Locally variable	High	Moderate	Low	Very low
(b) Coastal zone					
Delta	Widespread	Very high	High	Low	Very low
Esturine	Locally variable	High	Moderate	Low	Very low
Beach					
High wave energy	Widespread	Moderate	Low	Very low	Very low
Low wave energy	Widespread	High	Moderate	Low	Very low
Lagoonal	Locally variable	High	Moderate	Low	Very low
Fore shore	Locally variable	High	Moderate	Low	Very low
(c) Artificial					
Uncompacted fill	Variable	Very high	-	-	-
Compacted fill	Variable	Low	-	-	-

and high shear resistance. About the only objectionable feature of gravelly soils is possibly high permeability. The all-around properties of gravelly soils become improved with small amounts of clay binder. Reduction of seepage in gravels is best accomplished by adding clay as a binder. However, excess clay makes the soil revert to a less stable condition. If piping is found to be a problem (fine soil carried by water through a coarse soil), a filter or zone of intermediate size soil is necessary to prevent the movement of soil particles.

4.6 Rock

Solid rock is only mentioned here to complete the circle of earth types. It is usually considered to be the best type of foundation and has little reason to be questioned. However, rock varies in strength and sometimes is reported as bedrock when it is only a formation of consolidated clay. Also, there can be problems with some solid rocks, such as limestone or dolomite, which under certain ground water conditions, may dissolve and form sinkholes. Rocks that contain soluble salts, such as gypsum, are unsuitable for water retention structures.

4.7 Shales

Some shales and dense clays lose their strength from drying and cracking and subsequent wetting. This behavior is sometimes termed "slaking." Some shale soils also lose strength and disaggregate when exposed to water. Therefore, it is desirable that such materials be protected from drying and deterioration before the overlying part of a structure is built. Protection can be provided by a coating (asphalt emulsion, shotcrete, plastic, etc.) or moist soil. Also, some fresh shales have slicken slides that may be a stability problem when they are cut into (refer to section 4.2.1).

Figure 54 shows air slaking of shale. This is a block sample cut from fresh shale that was not exposed to air. The sample was cut open. The figure shows the effects of air on the shale. Note the blockiness of the material. This could

Figure 54.–A block sample of fresh shale allowed to air slake.

have been slicken slides that dried out. Since shale is a heavily overconsolidated clay, there is a good chance ot rebound and swell (sec. 4.2.3).

4.8 Loessial Soil

If a loaded Loess deposit is wetted, it rapidly consolidates and the structure constructed on it settles. Because of this property, Loess may be a dangerous foundation material if brought into contact with water. Failures of smaller structures on Loess that become saturated are numerous. In a spectacular case of an overnight settling and cracking of a house, the accident was caused by the discharge of water from a hose forgotten on the lawn.

Another difficulty with Loess is its ready ability to "pipe" under the action of water. If water starts to leak from an excavation or a canal, it forms a patch inside the Loess mass, which gradually progresses and widens, until a failure occurs. Similar accidents may also take place in the case of steel pipes placed in Loess. Water

99

finds its way around the pipe, and cavities as large as 9 feet in diameter have been reported. Presumably, an accident of this sort can be prevented by a careful placing of the pipe and the backfill around it.

The Loess settlement problem does not appear too serious if a concrete structure is built on a foundation that is not in contact with water. Such structures are, for instance, the footings of the towers of transmission lines or similar installations. Such towers, however, should not be placed in local depressions, which would permit water to accumulate.

Remolding of several upper feet of Loess at the surface and careful recompaction of the remolded material may create a reliable platform for building footings. Compacted Loessial soil canal linings have proved to be entirely satisfactory. If properly compacted, the Loess acquires a considerable shearing strength and resistance to erosion.

Table 33 can be used to determine if the Loess based on the in-place density needs treatment to support a structure. The sand cone density

method and the hollow stem auger with liner have been the proven methods to obtain the in-place density of Loess. Block samples can be tested for collapse. Other methods of sampling have a tendency to compact the sample.

Excavation in Loess usually is not difficult because of the capacity of Loess material to stand on almost vertical slopes.

4.9 Organic Soils

Organic soils should not be used as a construction material, because they decompose, compress, and will allow water to pass readily through them where this is not desired.

4.10 Dispersive Soils

Certain clay soils are susceptible to erosional and piping failure. These clays are called "dispersive" soils due to their tendency to disperse or deflocculate in water. The tendency for dispersion depends on several variables, such as mineralogy and chemistry of the clay

Table 33.—Settlement upon saturation vs. natural density: Loessial soils from Kansas and Nebraska

D_R	Density		Settlement potential	Surface loading
	lb/ft^3	g/cm^3		
Loose	< 80	< 1.28	Highly susceptible	Little or none
Medium dense	80-90	1.28-1.44	Moderately susceptible	Loaded
Dense	> 90	> 1.44	Slight, provides capable support	Ordinary structures

Notes:
1. For earth dams and high canal embankments, γ = 85 lb/ft^3 (1.36 g/cm^3) has been used as the division between high-density loess requiring no foundation treatment, and low density Loess requiring treatment.
2. Moisture contents above 20% will generally result in full settlement under load.

and dissolved water in the soil pores, with the primary factor being an abundance of sodium cations within the pore fluid. Numerous failures have occurred in water-retaining embankments constructed with dispersive clays. The failures are usually initiated with cracking, allowing for concentrated seepage paths and progressive erosion and piping.

Dispersive clays can sometimes be identified by surface indications, such as unusual erosional patterns with tunnels and deep gullies, together with excessive turbidity in any storage water. Areas of poor crop growth may also indicate highly saline soils, which in many cases, may be dispersive. Lack of surface evidence does not in itself preclude the presence of dispersive clays. Dispersive clays cannot be identified by standard laboratory index tests, such as grain size analysis or Atterberg Limits. Special tests, such as the "crumb," double hydrometer, pinhole, and soil chemistry tests, can be used to identify dispersive soils. The easiest method for field evaluation is the crumb test, where a 15-mm cube of soil at natural moisture content is placed in 250 mL of distilled water. If a colloidal cloud forms easily, a positive reaction is obtained, and the soil is most likely dispersive. The reader is cautioned that the crumb test is not completely definitive of dispersion potential. A dispersive soil may sometimes give a nondispersive reaction in the crumb test. Soils containing kaolinite with known field dispersion problems have shown nondispersive reactions in the crumb test. However, if the crumb test indicates dispersion, the soil is probably dispersive. The pinhole test best represents the physical processes of progressive dispersion and erosion. For more information on dispersive soils, the reader is referred to Sherard [33].

Dispersive clay soil use in constructing water-retaining embankments should be avoided.

Appropriate protection with the use of graded filters will allow use of dispersive soil. Normally, graded concrete sand will provide protection against migration. Dispersive soil should also be avoided for use as compacted clay lining without a graded protective cover. An alternative remedy for use of dispersive soil is by stabilization with lime or other additives that can neutralize excess sodium salts.

Dispersive clay is a special class of erosive soil reserved for clays with high colloid contents and high plasticity indices. Soils with colloid contents less than 15 percent and plasticity less than 13 percent are erosive. During investigations, if the crumb test indicates colloidal behavior, the soil can be checked further by pinhole testing [2, USBR 5410]. If dispersive clays are identified and they require treatment, the design team can assume that 1 percent of either cement or lime will be required. For construction purposes, specify 3 percent, since 1 percent is difficult to measure and control.

4.11 Erosive Soils

Erosion is generally caused by poor drainage patterns. Highly erosive soils are primarily sands and silts with little to no plasticity. An example of erosion problems can be seen on figure 55, wherein the drainage is across the road and then down the canal side slope.

In some cases, one fix may create another problem as shown on figure 56. Local gravelly material was use as the canal lining. This material is very permeable, and a geomembrane was installed to stop the loss of water from the canal. As a result of poor drainage patterns, the coarse material slid along the geomembrane. As can be seen, water drained down the upper berm and across the road.

Figure 55.—Erosion caused by drainage across an O&M road.

Figure 56.—Gravel-lined canal—water drained down an open slope, across a road, and along the lining under the gravel.

Figure 57.—Undercut erosion due to water flow.

If dispersive clays are present, or erosion is reoccurring or occurs in turbulence zones as shown on figure 57, either gravel or rock protective covers should be used, or velocity (tractive) forces should be limited. The figure shows how the canal lining was undercut up to the gravel and rock protection.

4.12 Slope Stability

An understanding of geology, hydrology, and soil properties is central to applying slope stability principles properly. Factors that affect slope stability are gravity, water, earth materials, and triggering events. Slides may occur in almost every conceivable manner, slowly of suddenly, and with or without any apparent provocation. Usually, slides are due to excavation or to undercutting the toe of an existing slope. In some instances, slides are caused by a gradual disintegration of the structure of the soil, starting at hairline cracks, which subdivide the soil in angular fragments, or as water content increases, the material can turn into a slurry and flow.

In most applications, the primary purpose of slope stability analysis is to contribute to the safe and economic design of excavation, embankments, landfills, and spoil heaps. Slope stability evaluations are concerned with identifying critical geological, material, environmental, and economic parameters that will affect the project, as well as understanding the nature, magnitude, and frequency of potential slope problems.

The objectives of slope stability analyses are to (1) understand the development and form of natural slopes, and the processes responsible for natural features, (2) assess the stability of slopes under short term and long term conditions,

(3) assess the possibility of landslides involving natural and existing engineered slopes, (4) analyze landslides and understand failure mechanisms and the influence of environmental factors, (5) enable the redesign of failed slopes and the planning and design of preventive and remedial measures, where necessary, and (6) study the effect of seismic loadings on slopes and embankments.

Figure 58 shows a slide along a road cut. The slide was caused where moisture entered the low density material. It can be seen that the slope was not cut back as far as the slope farther up the road.

One rough approach to the stability analysis without having soil strength parameters would be to perform an analysis on the slope that did not fail and back-calculate the soil strength parameters for a safety factor of 1. The in-place density of material would be needed to narrow down the unknown parameters. The results of the analysis would give an estimate of the strength parameter. This could be used for comparison of typical parameters for this type of material. A stability analysis could be performed using these strength parameter to determine how much to cut back the slope. This approach should only be performed by an experienced geotechnical engineer familiar with local soils.

Figure 59 shows a slope failure of a canal through a 90-foot cut. This is shown, to discuss several causes of failure. This can happen in any size cut. The slope failed on a curve. Lateral support diminishes on a curve. The waste material from the cut was piled on top of the slope, adding weight. Two slides occurred. As the slide progressed, the material in the center continued to slide, whereas the edges stopped sliding. The water in the canal did not

Figure 58.—A slide occurred where moisture entered a low density material.

Figure 59.—A second slide developed in the center of the first slide.

come from filling it, but from ground water draining into the canal prism.

Figure 60 shows a sample of material that was taken in the slide area. The material is clay with sand lenses. The sand lenses provided a drainage path. When the canal was cut, these lenses were pinched off. This caused the water pore pressure to build up and saturate the slope. When the clay became saturated, it lost its cohesive strength and contributed to the failure.

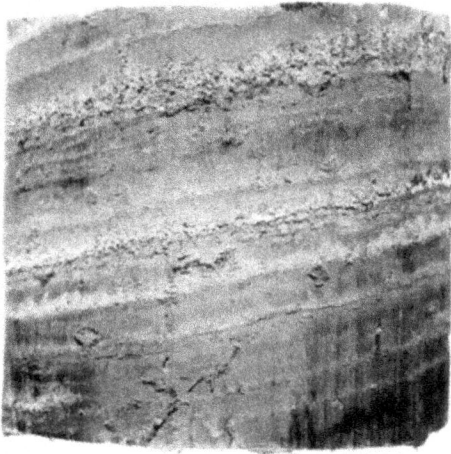

Figure 60.—A sample of material taken in the slide area of the canal. The material is clay with sand lenses.

Figure 61 shows how the slide was fixed. This is looking from the opposite direction. As a result of the slope stability analysis, the slopes were cut back, and the waste pile was reduced and spread over a larger area. Horizontal drains were drilled both above and below the lower O&M road. It was observed that the ground water was draining through these horizontal drains. This fix required purchase of additional right-of-way.

It also can be seen the original cut slopes are not stable. On the right side of canal, as shown on figure 59, slopes failed between the two roads.

4.13 Cuts

Cut slopes are an important feature on any project. The intent in a slope design is to determine a height and inclination that is economical to construct and that will remain stable for a reasonable lifespan. The design is influenced by the purposes of the cut, geological conditions, in situ material properties, seepage pressures, construction methods, and potential occurrence of natural phenomena, such as heavy precipitation, flooding, erosion, freezing, and earthquakes.

Cuts in clay shales and stiff, fissured clays are a concern. Refer to sections 4.2.1 and 4.6 for precautions.

Figure 61.—The slide was fixed by cutting back the slopes and reducing the waste pile and spreading it over a larger area. Horizontal drains were drilled both above and below the lower O&M road.

Steep cuts often are necessary because of right-of-way and property line constraints. The design must consider measures that will prevent immediate and sudden failure as well as protect the slope over the long term, unless the slope is cut for temporary reasons only. In some situations, cut stability at the end of construction may be critical design consideration. Conversely, cut slopes, although stable in the short term, can fail many years later without much warning.

Figure 62 shows a 100-foot cut through Loessial soil. The slope was cut a ¾ to 1. A layer of weak material popped out as a result of exposure and wind action through the canal prism.

Observation of long term cuts in the area will give a indication of how much to cut the material.

4.14 Fills

Fill slopes generally involve compacted soils. The engineering properties of materials used in these structures are controlled by the borrow source grain size distribution, the methods of construction, and the degree of compaction. In general, embankment slopes are designed using shear strength parameters obtained from tests on samples of the proposed material compacted to the design density. The stability analyses of embankments and fills do not usually involve the same difficulties and uncertainties as natural slopes and cuts, because borrow materials are preselected and processed. The main concern will be the underlying material the fill is placed on.

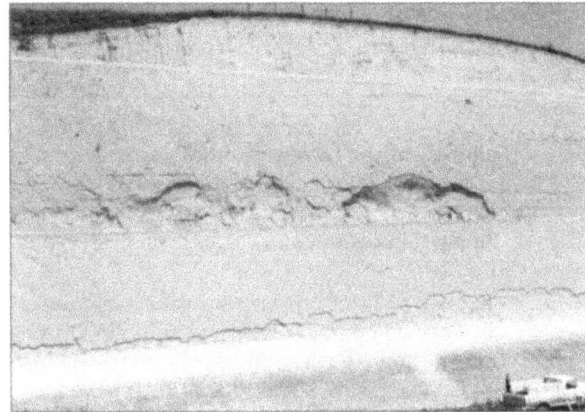

Figure 62.–Popouts in a ¾:1 cut in Loess, caused by weak material.

4.15 Uncontrolled Seepage and Piping

4.15.1 Uncontrolled Seepage

Seepage must be controlled on hydraulic conveyance systems, such as canals and diversion structures. Examples of failures by uncontrolled seepage are summarized on table 34 [34]. Uncontrolled seepage can weaken soil and initiate failures. Depending on the soils, drainage elements will be required for numerous features. For example, drains might be needed for embankments, and free draining backfill is required for retaining walls. This might necessitate location of free draining materials as part of the investigation. If small quantities are required, local concrete aggregate processing plants have good free draining materials.

For water-retaining embankments, soil is generally required to contain 25 percent fines to be impervious.

Table 34.– Examples of the consequences of uncontrolled seepage [34]

Category 1 Failures caused by migration of particles to free exits or into coarse openings	Category 2 Failures caused by uncontrolled saturation and seepage forces
1. Piping failures of dams, levees, and reservoirs, caused by: 　a. Lack of filter protection 　b. Poor eompaction along conduits, in foundation trenches, etc. 　c. Gopher holes, rotted roots, rotted wood, etc. 　d. Filters or drains with pores so large soil can wash through 　e. Open seams or joints in, rocks in darn foundations or abutments 　f. Open-work gravel and other coarse strata in foundations or abutments 　g. Cracks in rigid drains, reservoir linings, dam cores, etc. caused by earth movements or other causes 　h. Miscellaneous man-made or natural imperfections 2. Clogging of coarse drains, including French drains	1. Most landslides, including those in highway or other cut slopes, reservoir slopes, etc., caused by saturation 2. Deterioration and failure of road beds caused by insufficient structural drainage 3. Highway and other fill foundation failures caused by trapped ground-water 4. Earth embankment and foundation failures caused by excess pore pressures 5. Retaining wall failures caused by unrelieved hydrostatic pressures 6. Canal linings, basement and spillway slabs. uplifted by unrelieved pressures 7. Drydock failures caused by unrelieved uplift pressures 8. Dam and slope failures caused by excessive seepage forces or uplift pressures 9. Most liquefaction failures of dams and slopes caused by earthquake shocks

4.15.2 Piping

Piping is the phenomenon of internal soil erosion within a water-retaining structure or its foundation. Causes of piping failures can be lack of filter protection; poor compaction along conduits in foundation trenches; animal holes, rotted roots, rotted woods; filters or drains with pores so large soil can wash through; open seams or joints in rocks in foundations; open-work gravel and other coarse strata in foundations; cracks in rigid drains, reservoir linings, and dam cores, caused by earth movements or other causes; or miscellaneous manmade or natural imperfections.

Certain soils are more susceptible to piping failures. For example, nonplastic sand and silty sands are more susceptible to piping. Clay soils,

except dispersive clays, are less susceptible. The piping process originates largely due to the presence of a high exit gradient. Once the pipe is initiated, it progresses upstream, if enough water is flowing to carry the eroded soil to the exit point. As a result, the piping process continues in an upstream direction, and the average hydraulic gradient to the pipe may be increased as the pipe grows.

Structures subject to underseepage and side seepage should be equipped with sufficient cutoff to prevent seepage failures or designed with sufficient bankment width to prevent seepage from daylighting. Water-retaining embankments can be assumed to crack, especially in critical areas, such as cut/fill transitions. In these cases, one must consider use of filter zones to prevent piping. Again, filter sand can also be obtained from local aggregate suppliers. C-33 concrete sand will filter almost all soils.

Figure 63 shows a canal drop structure that failed by side seepage. The soil was silty sand (SM),which is highly erosive. The seepage path along the side of the structure was not long enough to prevent piping. The solution to this problem would be to add cutoff collars to lengthen the seepage path or to use clay backfill.

4.15.3 Filters and Drains

Filters and drains are used to prevent piping and reduce hydrostatic uplift pressures. The material must be free draining, but at the same time, must be able to dissipate relatively high hydraulic heads without movement of either the filter material or the protected soil. Often, a single layer of material will be inadequate, and a two-stage filter should be designed. Fine sand, silt, or clay in the pervious material is objectionable; processing by washing or screening is often required to produce acceptable material from most natural deposits.

Although the quantity of pervious materials required for filters and drains is usually small, quality requirements are high. Grading requirements will be different; filter materials are commonly secured economically from sources acceptable for concrete aggregate. Particle shape of pervious material is not as critical; processed concrete aggregates rejected for shape can usually be used to construct drainage blankets and drains, if suitable gradation and adequate permeability are maintained. However, minerals contained in pervious materials should be evaluated for potential degradation as water percolates through the filter. Likewise, attention should be given to soundness and durability of particles to be sure no significant change occurs in gradation due to particle breakdown as the material is compacted.

Figure 63.—A canal drop structure that failed by side seepage.

ASTM C-33 concrete sand can serve as a protective filter material for many materials. Figure 64 shows filter criteria using the concrete sand. Reclamation has a filter design standard for embankment dams that should be consulted [35].

Recently, geosynthetic materials (geotextiles, geonets, and geocomposites) have gained wide acceptance for use as filters and drains in civil engineering works.

4.16 Collapse and Subsidence

Collapse can result from the sudden settlement of a foundation upon wetting. Soils above the ground water level that are of low density in their natural dry condition, when they become thoroughly wetted, can collapse appreciably. The low density condition is particularly severe in very arid portions of the western United States, where these types of soils were usually deposited as Loess or as quickly deposited materials on the outer limits of alluvial fans. More than 3 feet of settlement is common in widespread areas of collapsible/low density soils. Figure 65 shows differential settlement of

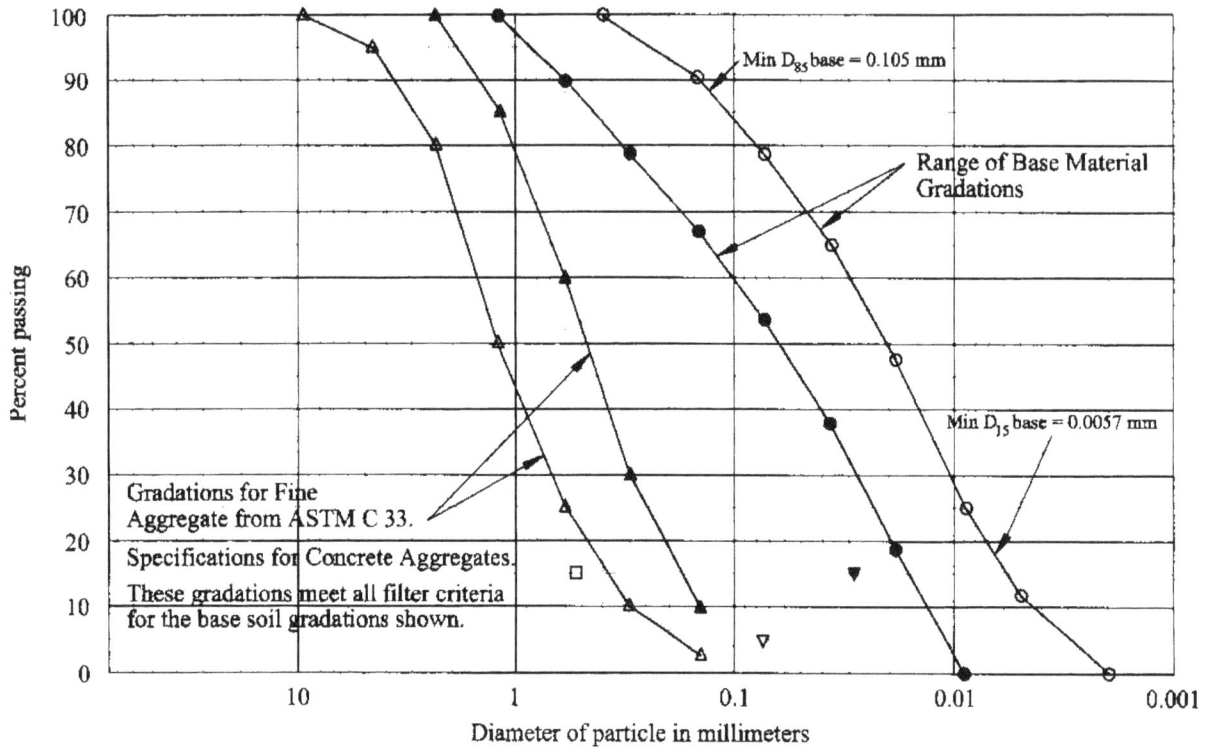

Plot symbol	Gradation ranges
O	Base material - finest
●	Base material - coarsest
△	Filter material - coarse limit of ASTM C 33 Fine Aggregate
▲	Filter material - fine limit of ASTM C 33 Fine Aggregate

Plot symbol	Filter design requirements
▽	Limit of fines in filter = 5% passing 0.075 mm
■	Maximum particle in filter = 100% passing 75 mm
▼	Permeability criteria : D_{15} filter > or = 5* D_{15} base D_{15} filter > of = 0.029 mm
□	Stability criteria : D_{15} filter < or = 5 * D_{85} base D_{15} filter < of = 0.525 mm

Figure 64.—Filter gradations.

Figure 65.—This structure was placed on uncompacted material, causing differential settlement of the structure.

a structure placed on uncompacted fill and low density Loessial soil.

Subsidence due to long term fluid withdrawal is also an area of potential settlement concern. Generally, subsidence occurs at such a slow rate that the foundations are not immediately affected. However, the long term use of a structure, such as a canal, can be severely impacted.

Explorations and analyses need to be performed in areas of potentially collapsible/low density

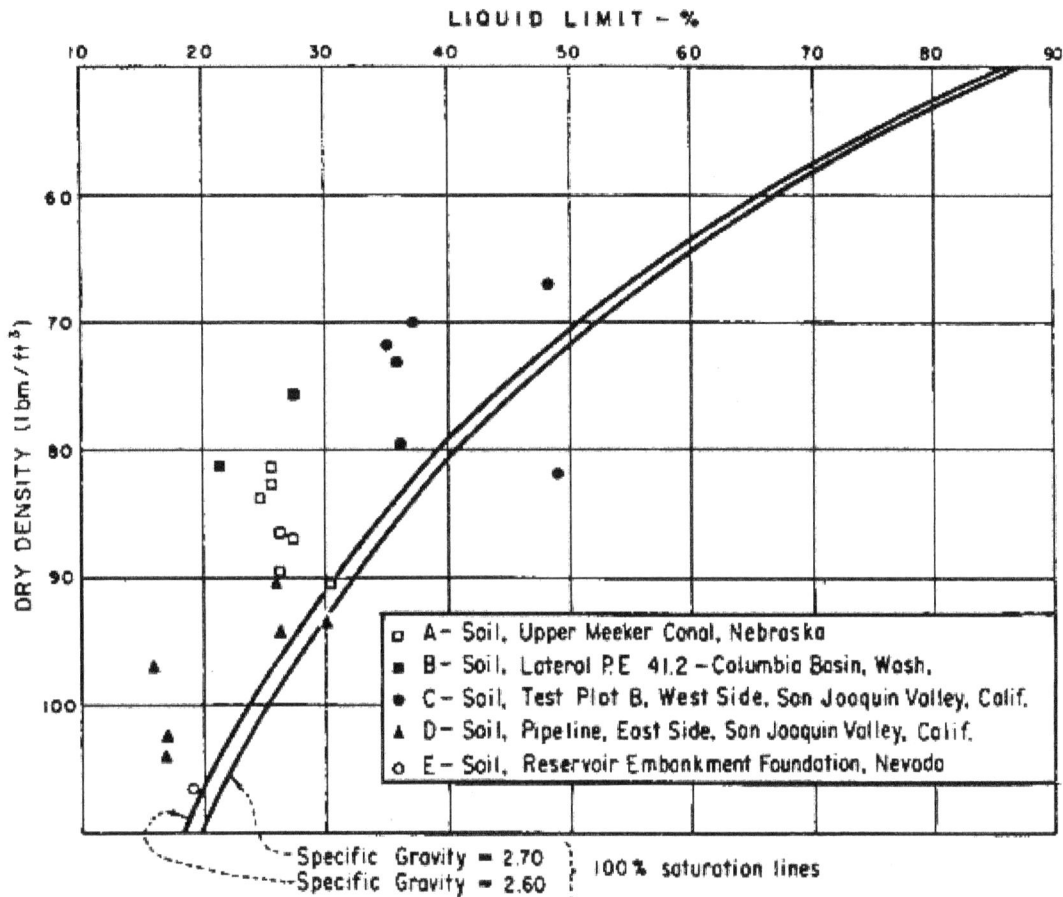

Figure 66.—Assessing collapse potential of various soils using dry density and liquid limit.

soils to determine the aerial extent, depths, and percentages of settlement that the soils will exhibit.

By obtaining the in-place density and liquid limit laboratory tests of the material, an assessment of collapse potential can be made using figure 66.

Also, collapse potential can be determined from one-dimensional consolidation tests by observing the change in void ratio when wetting the test specimen under the structure's design load. This is shown on figure 67.

Figure 68 shows the effect of prewetting on consolidation of Loessial soils from the Missouri River Basin. The void ratio decreases drastically when the low-natural-water-content soil is wetted.

4.17 Frost Heave

Frost action tends to lift soils by moving water from below (by capillary action) into the freezing zone and forming ice lenses in the soil. This usually takes place in cold climates, when the water table is within a few feet of the ground surface and the soils are fine grained. Frost heave not only lifts structures but also

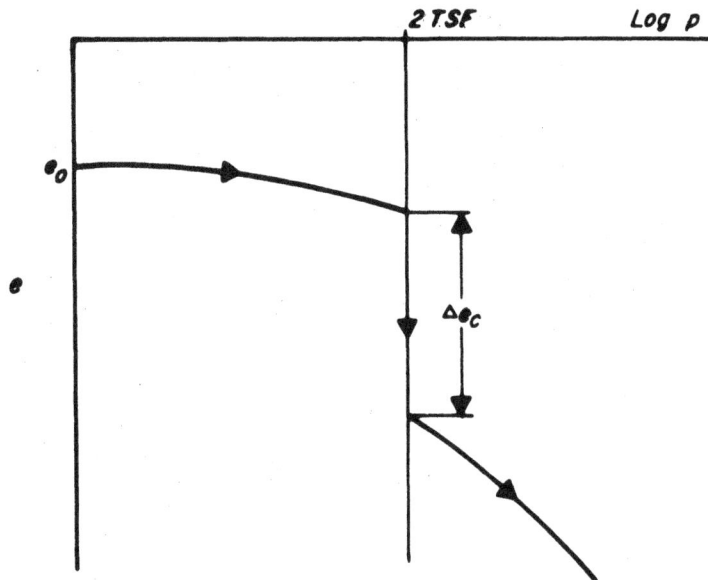

Collapse Potential (CP) is defined as:

$$CP = \frac{\Delta e_c}{1 + e_0} \qquad \text{or} \qquad CP = \frac{\Delta Hc}{Ho}$$

Δe_c = Change in void ratio upon wetting ΔHc = Change in height upon wetting

e_0 = Natural void ratio Ho = Initial height

Collapse Potential Values

CP	Severity of Problem
0-1%	No problem
1-5%	Moderate trouble
5-10%	Trouble
10-20%	Severe Trouble
20%	Very Severe Trouble

Figure 67.—Typical collapse potential test results [17].

permits piping after the ice lenses melt. Extensive repair work may be necessary after the spring thaw. Where necessary, frost heave can be reduced principally by (1) burying the footings below the frost line, (2) placing earth pads (sand or gravel) beneath the structure, (3) using polystyrene or some other form of insulation between the soil and concrete lining, or (4) draining the water to prevent buildup of ice lenses. Clean, granular soils, rather than silts and clays, are most effective in protecting against uplift due to frost action, because the larger space between particles prevents the rise of water due to capillary action and the buildup of ice lenses. However, coarse, granular soils have less value as insulation than fine-grained soils.

110

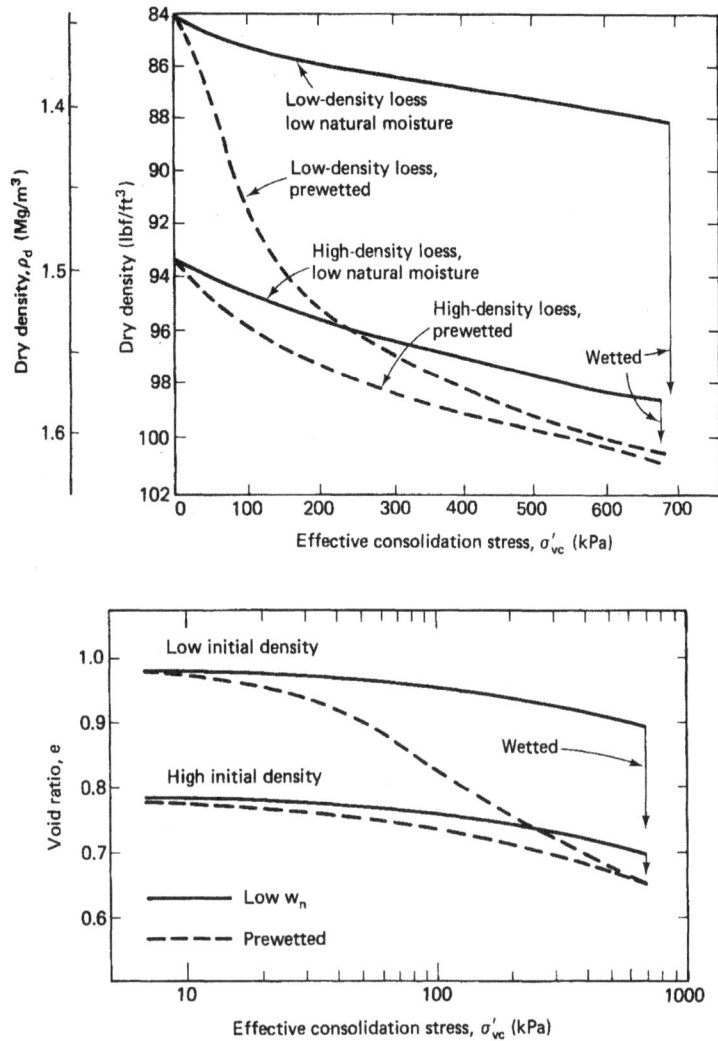

Figure 68.—Loessial soils from the Missouri River Basin, showing effect of prewetting on consolidation. Note the drastic reduction in the void ratio when the low natural water content soil is wetted [13].

Freezing also affects the properties of soils when there are no sources of water accept water in the soil voids. When construction on an earth dam is shut down for the winter; freezing near the surface causes an increase in moisture content and a decrease in density near the surface. Snow acts as insulation and affects the depth of frost penetration. Before construction is resumed in the spring, and after frost is out of the ground, density tests need to be performed to make sure that the specification requirements for density and moisture are met.

Although frost action also affects the density and moisture of compacted earth linings after water is taken out of a canal, the overall effects that would cause serious damage and an increase in seepage do not seem to occur where the water table is not near the lining. Where the water table is within a few feet of the lining and especially when the canal is on a side hill where ground water can drain into the lining, as shown on figures 69 and 70, then ice lenses will build up in the lining. This will reduce soil density and cause increased seepage through the lining.

111

Figure 69.–Frost heave of a canal lining with drainage toward the canal.

Figure 70.–Canal lining cracking due to frost heave.

Alignment of a canal in an east-west direction will generally have frost heave problems on the south side due to lack of direct sunlight on the lining.

Figure 71 illustrates results of frost heave laboratory tests. The figure shows the heave rate of expansion related to the percentage of mass finer 0.02 mm. Based on the heave rate, a frost susceptibility classification is derived.

Frost heaving susceptibility is also shown in table 35 for soils in a loose to medium-compact state. This table identifies the soil using D_{10}, the grain size diameter for which 10 percent of the material is finer.

4.18 Soil Modification

A modified soil is a mixture of soil, water, and a small amount of an additive. Soil stabilization is the chemical or mechanical treatment of soil to improve its engineering properties. Chemically stabilized soils consist of soil and a small amount of additive, such as cement, fly-ash, or lime. The additive is mixed with the soil, and the mixture is used in compacted fills, linings, or blankets. Quality and uniformity of the admixture and the uniformity of moisture are closely controlled to produce a high quality end product.

Both methods of soil additives discussed here require laboratory testing programs. This is to identify if the soil is suitable for the additive. The second part of the laboratory programs is to determine the minimum percentage of additive to use that will satisfy the structure's design requirements.

4.18.1 Soil-Cement

Soil-cement has many uses, such as slope protection for dams and other embankments, and linings for highway ditches, canals, other channels, reservoirs, and lagoons. Soil-cement's low cost, ease of construction, and convenient utilization of local or in-place sandy soil makes such applications economical, practical, and environmentally attractive. Soil-cement becomes cost effective when used for large areas.

Soils suitable for soil cement are sands and silty sands with up to 30 percent fines. Gravelly sand can also be used, but the construction methods would change. The soil deposits should have very low clay (clay ball) content. The typical range of cement content is 8 to

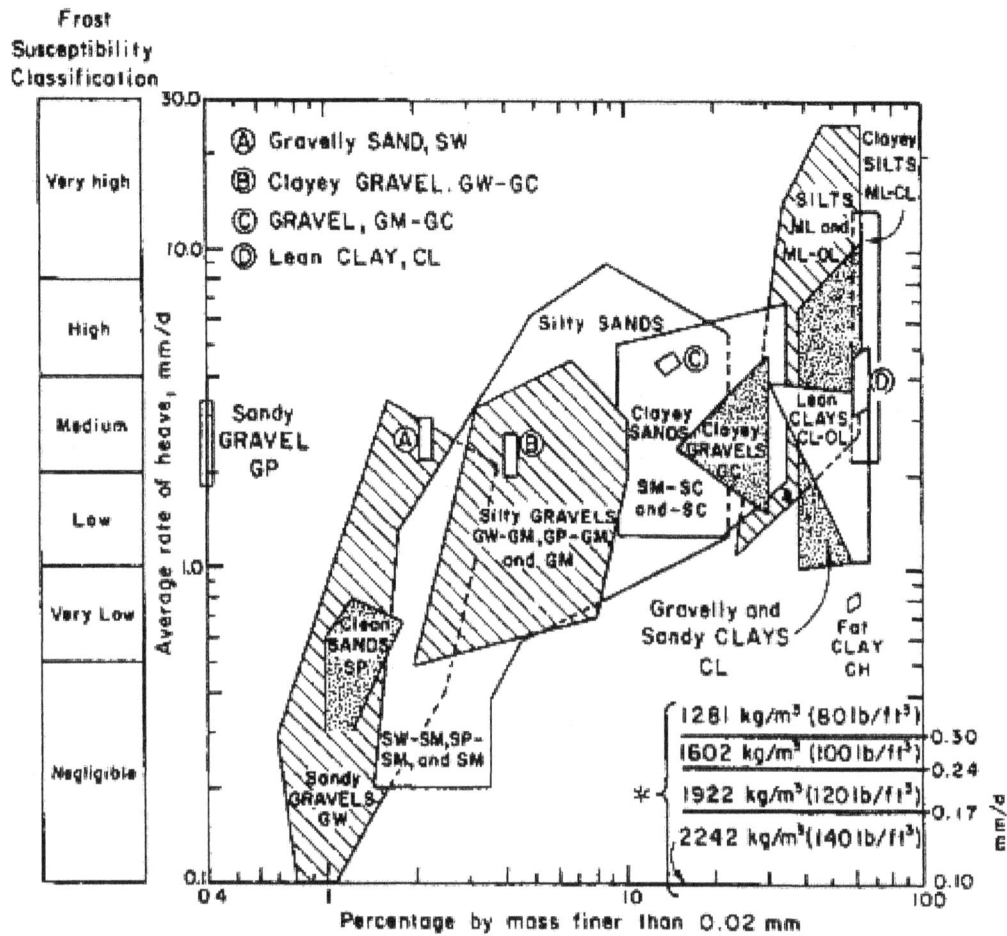

Figure 71.—Frost susceptibility classification by percentage of mass finer than 0.02 mm.

12 percent. If lab testing is too costly, the Portland Cement Association has design guides for soil cement. For estimating purposes, assume 10 to 12 percent cement will be required.

For slope protection, the layer should be (1) formed into a homogeneous, dense, permanently cemented mass that fulfills the requirements for compressive strength (typically 850 lb/in² at 28 days), (2) in intimate contact with earth slopes, abutments, or concrete structures, (3) durable and resistant to "wetting and drying" and "freezing and thawing" actions of water, and (4) stable with respect to the

structure and of sufficient thickness (mass) to resist displacement and uplift. Figure 72 shows what wave action can do to slope facing.

The wave damaged and wave washed areas were caused by inadequate bonding of the soil-cement layers. Several studies have since been performed to identify methods for enhancing bond between layers. Currently, the most promising method investigated is to apply a water-cement slurry to a layer just before placing the overlying layer.

Table 35. —Tentative criteria for rating soils with regard to drainage, capillarity, and frost heaving characteristics[1]

	"Trace fine sand"	"Trace silt"	"Little silt" (coarse and fine)	"Some fine silt" "Little clayey silt" (fissured clay sails)	"Some clayey silt" (clay soils dominating)
Fineness identification [2]	"Trace fine sand"	"Trace silt"	"Little silt" (coarse and fine)	"Some fine silt" "Little clayey silt" (fissured clay sails)	"Some clayey silt" (clay soils dominating)
Approx. effective size, D_{10}, mm [3]	0.4 0.2	0.2 0.074	0.074 0.02	0.02 0.01	0.01
Drainage	Free drainage under gravity excellent	Drainage by gravity good	Drainage good to fair	Drains slowly, fair to poor	Poor to impervious
Approx. range of k, cm/s	0.5 0.2	0.10 0.04	0.020 0.006	0.0010 0.0004	0.0002 0.0001
	◄—— Deep wells	——— Well points successful ———►			
Capillarity Approx. rise in feet, H_0	Negligible 0.5	Slight 1.0	Moderate 1.5 3.0	Moderate to high 7.0 10.0	High 15.0 25.0
Frost heaving susceptibility	Nonfrost-heaving	Slight	Moderate to objectionable	Objectionable	Objectionable to moderate
Groundwater within 6 ft or $H_0/2$					

[1] Criteria for soils in a loose to medium-compact state.
[2] Fineness classification is in accordance with the ASCE Classification System
[3] Hazen's D_{10}. The grain size for which 10% of the material is finer.

114

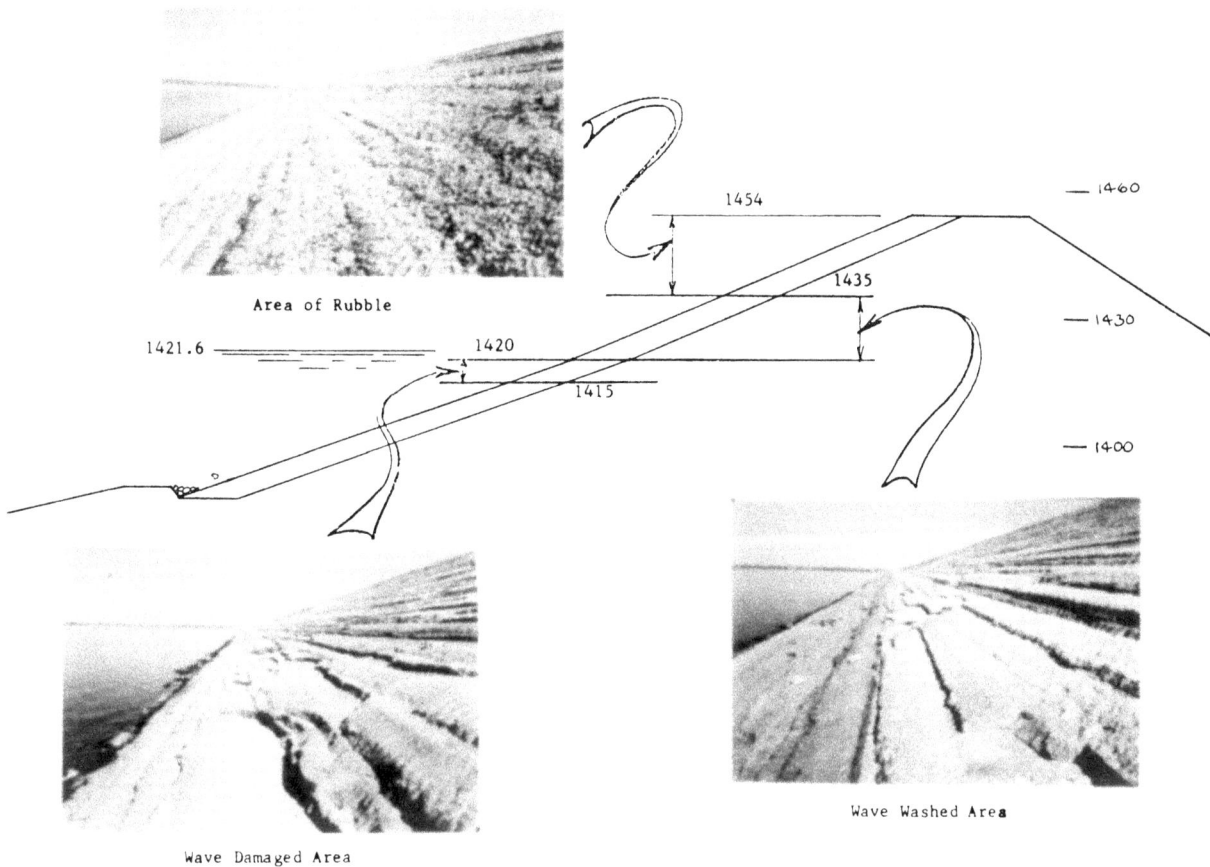

Figure 72.—Wave damage on a soil-cement slope facing.

4.18.2 Soil-Cement Slurry

Soil-cement mixtures can also be used as a slurry and for rapid pipe embedment. This is a cement-stabilized soil, consisting of amixture of soil and cement with sufficient water to form a material with the consistency of a thick liquid that will flow easily and can be pumped without segregation. Soil-cement slurry has many other names, including flowable fill, and controlled low strength materials (CLSM). Most soil-cement slurry is now procuced at the batch plant. In flowable fill, the soil aggregate is typically sand and fine gravel aggregate, but a wide variety of materials can be used. Sands with up to 30 percent nonplastic or slightly plastic fines are best. Soil-cement slurry has

been used for pipe bedding. Even though materials from the trench excavation may be used, locating the borrow areas along the pipeline alignment is generally more economical and usually results in a better controlled and more uniform product. If soils onsite are not acceptable, flowable fill slurry can be obtained from ready mix plants. Soil-cement slurry pipe embedment must not be too strong, so the material can be excavated if repair is needed. A 28-day strength of 50 to 100 lb/in^2 is about right. More information on soil cement slurry for pipe construction can be found in Geteochnical Branch Training Manual No. 7 [12].

115

4.18.3 Lime Stabilization

Fine-grained soils exhibit improved plasticity, workability, and volume change characteristics when mixed with lime; however, not all soils exhibit improved strength, stress-strain, and fatigue characteristics. Properties of soil-lime mixtures depend on many variables. Soil type, lime type, lime percentage, and placing and curing conditions, including time, temperature, and moisture are the most important variables.

Two types of clay soils have been encountered that cause special difficulties for some structures. The first is expansive clay, and the second is dispersive clay. Lime treatment of both types of clays has been found to be an effective stabilization method.

Adding lime to soil has two major effects:

- The first effect is improving the soil workability and also increasing the soil strength. This is immediate and results from the following reactions of the lime with the soil: (1) an immediate reduction in plasticity, where the liquid limit of the soil is decreased and the plastic limit increased, thus reducing the plasticity index of the soil, (2) the finer clay-size particles agglomerate to form larger particles, (3) the large particles (clay clods) disintegrate to form smaller particles, and (4) a drying effect takes place due to the absorption of moisture for hydration of the lime, which reduces the moisture content of the soil. The result of these reactions is to make the material more workable and more friable or siltlike in texture. This eliminates the construction problems inherent in using wet, sticky, heavy clay.

- The second effect of adding lime to soil is a definite cementing action, with the strength of the compacted soil-lime increasing with time. The lime reacts chemically with the available silica and some alumina in the soil to form calcium silicates and aluminates.

If it is determined that lime treatment may be required, the design team can assume 3 to 4 percent lime by dry mass of soil. There are two types of lime, "quick" lime and hydrated lime. Hydrated lime contains water, and about 30 percent more lime than dry lime would be required to be added to obtain the same dry mass.

Certain clay soils contain sulfates that can adversely affect lime treatment. Sulfates react with the lime to form gypsum gels and actually cause the soil to swell more. For these soils, the lime requirement is normally doubled to 6 to 9 percent. The lime is added in two applications, and mellowing time is increased to 2 to 4 days. Prior to treatment of any clay with lime, sulfate testing should be performed.

Bibliography

[1] Bureau of Reclamation, *Earth Manual*, Part 1, 1998, Third Edition, Denver, CO.

[2] Bureau of Reclamation, *Earth Manual*, Part 2, 1990, Third Edition, Denver, CO.

[3] Bureau of Reclamation, *Engineering Geology Field Manual*, 1991, Denver, CO.

[4] Bureau of Reclamation, *Engineering Geology Office Manual*, 1988, Denver, CO.

[5] Bureau of Reclamation, *Ground Water Manual*, 1995, Denver, CO.

[6] Bureau of Reclamation, *Drainage Manual*, 1993, Denver, CO.

[7] Bureau of Reclamation, *Design of Small Dams*, 1987, Denver, CO.

[8] Bureau of Reclamation, *Design of Small Canal Structures*, 1978, Denver, CO.

[9] Howard, A.K., and Bara, J.P., *Lime Stabilization on Friant Kern Canal*, 1976, Report REC-ERC-76-20, Bureau of Reclamation, Denver, CO.

[10] Bureau of Reclamation, *Linings For Irrigation Canals*, 1963, First Edition, U.S. Government Printing Office, Washington, D.C.

[11] Howard, A.K., L. Kinney, and R. Fuerst, *Method for Prediction of Flexible Pipe Deflection*, Revised March 1, 2000, M25, Bureau of Reclamation, Denver, CO.

[12] Bureau of Reclamation, *Pipe Bedding and Backfill*, 1996, Geotechnical Branch Training Manual No. 7, Denver, CO.

[13] Holtz R.D. and W.D. Kovacs, *An Introduction to Geotechnical Engineering*, 1981, Prentice Hall, Inc., Englewood Cliffs, NJ.

[14] Hunt, Roy B., *Geotechnical Engineering Investigation Manual*, 1984, McGraw-Hill, New York, NY.

[15] Bureau of Reclamation, *Concrete Manual*, 1988, Part 1, Ninth Edition, U.S. Government Printing Office, Washington, D.C.

[16] Bureau of Reclamation, *Corrosion Considerations for Buried Metallic Water Pipe*, 2004, Technical Memorandum No. 8140-CC-2004, Water Conveyance Group, Technical Service Center, Denver, CO.

[17] Department of the Navy, *Foundations and Earth Structures*, 1982, Design Manual 7.2, NAVFAC DM-7.2, Naval Facilities Engineering Command, Aleaxndria, VA.

[18] Terzaghi K., and R.B. Peck, *Soil Mechanics in Engineering Practice*, 1967, Second Edition, John Wiley and Sons, New York, NY.

[19] Lambe, T.W., and R.V. Whitman, *Soil Mechnics*, 1969, John Wiley and Sons, New York, NY.

[20] Peck, R B, W. E. Hanson, and T.H. Thornburn, *Foundation Engineering*, 1974, John Wiley and Sons, New York, NY.

[21] Bureau of Reclamation, *Soil Classification Handbook*, 1986, Geotechnical Branch Training Manual No. 6, Denver, CO.

[22] *Geotechnical Engineering Techniques and Practices*, Roy E. Hunt, 1986, McGraw Hill Book Company, NY.

[23] Hough, B.K., *Basic Soils Engineering*, 1957, The Ronald Press Company, New York, NY.

[24] Bureau of Reclamation, *Seismic Design and Analysis*, 1987, Design Standards No. 13, "Embankment Dams," Chapter 13, Denver, CO.

[25] American Society for Testing and Materials, *Annual Book of Standards*, 1987, "Standard Method for Field Measurement of Soil Resistivity Using the Wenner Four Electrode Method," vol. 03.02, West Conshohocken, PA.

[26] Farrar, J., *Standard Penetration Test: Driller/Operators Guide*, 1999, Dam Safety Office Report DSO-98-17, Bureau of Reclamation, Denver, CO.

[27] American Society for Testing and Materials, *Annual Book of Standards*, 2001, "Standard Practice for Determining the Normalized Penetration Resistance of Sands for Liquefaction Resistance Evaluations," vol. 04.09, "Construction," West Conshohocken, PA.

[28] American Society for Testing and Materials, *Annual Book of Standards*, 2001, "Practice D 3550 on Thick Wall, Ring Lined, Split Barrel Drive Sampling of Soils," vol. 04.09, "Construction, Soil and Rock" West Conshohocken, PA.

[29] Lunne, T., P.K. Robertson, and J.J.M. Powell, *Cone Penetration Testing in Geotechnical Practice*, Blackie Academic and Professional, New York, NY.

[30] American Society for Testing and Materials, *Annual Book of Standards*, 2001, "Practice D6151 Using Hollow-Stem Augers for Geotechnical Exploration and Soil sampling," vol. 04.09, "Construction, Soil and Rock" West Conshohocken, PA.

[31] American Society for Testing and Materials, *Annual Book of Standards*, 2001, D 2113, "Standard Practice for Rock Core Drilling and Sampling of Rock for Site Investigation," vol. 04.08, "Soil and Rock" West Conshohocken, PA.

[32] Bieniawski, Z.T., *Engineering Rock Mass Classifications*, 1989, John Wiley and Sons, New York, NY.

[33] Sherard, J.L., and R.S. Decker, eds., *Dispersive Clays, Related Piping, and Erosion in Geotechnical Projects,*" 1977, STP 623, American Society for Testing and Materials, Philadelphia, PA, p. 101.

[34] Cedergren, H.R., *Seepage, Drainage, and Flownets*, 1988, Third Edition, John Wiley and Sons, New York, NY.

[35] Bureau of Reclamation, *Protective Filters*, 1994, Design Standards No. 13, "Embankment Dams," Chapter 5, Denver, CO.

[36] Church, Horace K., *Excavation Handbook*, 1981, McGraw-Hill Book Company, New York, NY.

Index

A

Adjacent structures, 19
Aggregate, 9, 17, 43, 105, 107, 115
Alignment, 4, 7, 9, 16, 115
Angle of internal friction of cohesionless soils, 48, 49
Atterberg limits, 1, 18, 26, 42, 43, 55, 66, 72, 78, 83, 93, 101
Augers
  bucket, 80–82
  hand, 72
  hollow stem, 83–84, 100
  power, 4, 5, 80–82

B

Backfill material, 9, 61–64
Background study, 11–20
Bearing capacity of structures, 49–51, 54
Bearing pressure
  allowed for rock and soil, 58
  nominal values allowed for spread foundations, 58
Boreholes, 18
Borrow areas, 4, 8, 9, 16, 17, 72, 80, 105, 115
Bucket augers, 70, 80–82
Buildings, 1, 3, 23

C

Canals, ix, 4, 5, 13, 91, 106, 4–5
  cracks in lining due to frost heave, 112
Cement, 7, 19, 23, 101, 112, 113, 115
Clays, 5, 17, 23, 35, 41, 42, 46, 57, 81, 93, 94, 106, 93–97
  compressible, 93–94
  expansive, 4, 94–97

firm, consolidated, 42, 50, 70, 71, 93, 94, 99, 93
Coefficient of consolidation, 45
Coefficient of friction, 49
Cohesionless soils, correlations between compactness and N, 74
Collapse, 107–9
Compacted soils, typical properties, 32
Compressibility, 18, 38–44
Compressible clays, 93, 94, 93–94
Compression index, typical values, 43
Compressive strength, 50, 51, 58, 68, 75, 84, 86, 113
Concrete, 4, 5, 6, 9, 18, 19, 23, 47, 64, 100, 101, 105, 107, 110, 113
Cone penetration testing, 71, 78–80
Consolidated clays, 93
Consolidation, 3, 18, 34, 40, 43–45, 52, 53, 67, 109, 111
Constrained modulus, 81
Corrosion, 18, 19, 23, 64
Costs, 6, 21, 22, 23, 24, 65, 72, 78, 84, 112
Cuts, 4, 17, 19, 104–5
  problem soils, 4, 17, 19, 104–5
  road cuts and fills, 19, 103, 19

D

Deformation modulus, 84
Density, 4, 6, 8, 17, 18, 25, 33, 37, 43, 44, 47, 49, 50, 58, 61, 66, 70, 72, 74, 75, 78, 79, 80, 87, 91, 94, 95, 97, 100, 103, 105, 107, 108, 109, 111
Depth, sampling, 23
Diamond rock coring, 84–86
Dispersive soils, 5, 101, 100–101
Distribution of principal soil deposits, 88
Dozer trenches, 71–72
Drains, 5, 8, 38, 107, 111, 114, 107

U, V, W

Appendix

Approximate Material Characteristics [36]

For about 100 years, commencing with Trautwine's pioneering handbook of 1882, *Civil Engineer's Pocketbook*, authoritative sources in the United States have been publishing tables of material characteristics. Generally speaking, these tables include specific gravities, weights in natural bed, swell factors from the natural bed or cut to the loose condition, weights in the loose condition, swell or shrink factors from the natural bed or cut to uncompacted fills or compacted embankments, and weights in uncompacted fills or compacted embankments. Engineers, both public and private, contractors, mining companies, machinery manufacturers, and writers of handbooks have contributed to this array of data.

The following table in this appendix is a summary of existing data, commencing with Trautwine's tables based on his own meticulous laboratory and field work, and ending with personal data gathered during the past 50 years. The table is necessarily based on properly interpreted and weighted averages. It is therefore not absolute for a specific case, and engineering experience and judgment will guide the user in its proper application. Prior to examination of the table, the reader is referred to these explanatory notes.

Materials

Rock materials are noted to be I, igneous; S, sedimentary; or M, metamorphic. Materials marked by asterisks are ores in the mineral or near mineral state, and the weights do not allow for the containing gangues of the ore body. The weight of the mineral is constant, with a set specific gravity, but the weight of the gangue, such as the associated earthy materials contained in quartz, rhyolite, schist, and feldspar, varies considerably with respect to the weight of the mineral. In mining the engineer must estimate the unit weight of the ore body and the weight of the contained mineral.

For example, hematite, the iron mineral, weighs 8560 lb/yd^3. Associated gangue, however, varies with respect to the hematite. Suppose that the mineral hematite samples 40 percent by weight of the ore and that the gangue, weighing 4000 lb/yd^3, samples 60 percent by weight of ore. Then 1000 lb of ore in the natural bed occupies a volume of

$$\frac{40\% \times 1000}{8560} + \frac{60\% \times 1000}{4000} = 0.197 \text{ yd}^3$$

The ore, then, weighs 1000/0.197=5080 lb/yd^3, as contrasted to the weight of 8560 lb/yd^3 for the contained mineral hematite. At this juncture it is well to explain that miners sometimes use the word *hematite* for both the mineral and the ore.

Specific Gravity

When the value for specific gravity is in parentheses, it is an *apparent specific gravity* because the material is not in the solid state.

Examples are gravel and rock-earth mixtures, which contain voids when in their natural bed.

Cubic Yard in Cut

The weight in the natural bed, or bank measurement, includes natural moisture. The average weight is subject to a maximum ±10 percent variation. Again, it is emphasized that ore weights are for the mineral only and not for an impure ore body containing gangue.

Cubic Yard in the Loose Condition

Percent swell from the natural bed to the loose condition is an average which is subject to a maximum 33 percent variation in both rock and earthy materials. Variations are multipliers and not percentages to be added to or subtracted from the given percent of swell. The swell factor of 67 percent, given for several rocks, is an average figure obtained from existing data for solid rock, and it has been applied to solidly bedded unweathered rocks for which no swell factors are available specifically. Percent swell factors for ores are in terms of the entire ore body rather than in terms of the contained mineral. Weights in the loose condition are averages, except when calculated on the basis of the aforementioned average 67 percent swell factor. All weights are subject to any adjusted value of the swell factor.

Cubic Yard in the Fill

In the table a cubic yard in a fill is a cubic yard in a compacted embankment. No values are given for ores in a fill as they are not construction materials. When they are in a fill, they are in a stockpile, and the values for a cubic yard in the loose condition are applicable. Percent swell or shrink from cut or natural bed to fill is an average, subject to a maximum

33 percent variation in both rock and earthy materials. Percentage variation is a multiplier.

It is absolutely necessary, especially in the case of rock materials, to distinguish between two methods of fill construction:

1. Natural or gravity compaction, which was common years ago before the development of compacting machinery, is little used now except in the building of waste fills and stockpiles of materials and ores. The swell and shrink factors from the cut or natural bed vary from 10 percent shrinkage for earthy materials to 67 percent swelling for rock materials. Because of different degrees of fragmentation in the cut and because of the wide variations of fill construction methods in natural or gravity compaction, no figures are tabulated.

2. Mechanical compaction by rollers, along with wetting of the fill, is today's accepted method for fill consolidation. The tabulated swell and shrink factors and weights are for this modern method. of fill compaction.

Two other influences affect swell and shrink factors and resultant weights. First, crawler-tractor-rippers produce better fragmentation and better grading of both rock and earthy formations in the cut. Second, the average so-called rock job really consists of a rock-earth mixture which in itself is pretty well graded.

These three factors, nature of materials, use of tractor-rippers, and modern compacting methods, have made possible the prevalent high densities of fills, densities not in accordance with some previously tabulated data for swell and shrink factors from cut to fill. In the case of construction materials the writer has used swell and shrink factors and weights, including moisture, resulting from average compaction methods.

It is a fact that certain friable rocks in weathered and parent rock zones have low swell factors from cut to fill. These rocks are really equivalent to rock-earth mixtures in their behavior during excavation and compaction. Rock swell factors are in terms of solid rock in the cut and do not include allowances for overlain residual and weathered rocks or for earthy and friable materials, all of which would reduce greatly the swell factor from cut to fill.

Approximate material characteristics

Material	sp gr	Cubic yards, in cut-weight, lb	Cubic yards, loose		Cubic yards in fill	
			Percent swell	Weight, lb	Swell or shrink, %	Weight, lb
Adobe, S	(1.91)	3230	35	2380	-10	3570
Andesite, I	2.94	4950	67	2970	33	3730
Asbestos	2.40	4040	67	2420		
Ashes, coal	(0.61)	1030	33	800	-50	2060
Asphaltum, S	1.28	2150	67	1390		
Asphalt rock, S	2.41	4050	62	2500		
Aragonite, calcium ore*	3.00	5050	67	3020		
Argentite, silver ore*	7.31	12300	67	7360		
Barite, barium ore*	4.48	7560	67	4520		
Basalt, I	2.94	4950	64	3020	36	3640
Bauxite, aluminum ore*	2.73	4420	50	2940		
Bentonite	1.60	2700	35	2000		
Biotite, mica ore*	2.88	4850	67	2900		
Borax, S	1.73	2920	75	1670		
Breccia, S	2.41	4050	33	3040	27	3190
Calcite, calcium ore*	2.67	4500	67	2700		
Caliche, S	(1.44)	2430	16	2100	-25	3200
Carnotite, uranium ore*	2.47	4150	50	2770		
Cassiterite, tin ore*	7.17	11380	67	6800		
Cement				2700		
Cerrusite, lead ore*	6.50	10970	67	6560		
Chalcocite, copper ore*	5.70	9600	67	5750		
Chalcopyrite, copper ore*	4.20	7060	67	4220		
Chalk, S	2.42	4060	50	2710	33	3050
Charcoal				1030		
Chat, mine tailings				2700		
Cinders	(0.76)	1280	33	960	-10	1420
Cinnabar, mercury ore*	8.10	13630	67	8170		
Clay, S:						
Dry	(1.91)	3220	35	2380	-10	3570
Damp	(1.99)	3350	40	2400	-10	3720
Clinker						2570
Coal, S:						
Anthracite	1.55	2610	70	1530		
Bituminous	1.35	2280	67	1370		
Coke	(0.51)	860	0	860		

Approximate material characteristics

Material	sp gr	Cubic yards, in cut-weight, lb	Cubic yards, loose		Cubic yards in fill	
			Percent swell	Weight, lb	Swell or shrink, %	Weight, lb
Colemanie, borax ore*	1.73	2920	75	1670		
Concrete:						
Stone	2.35	3960	72	2310	33	2910
Cyclopean	2.48	4180	72	2430	33	3150
Cinder	1.76	2970	72	1730	33	2240
Conglomerate, S	2.21	3720	33	2800	-8	4030
Decomposed rock:						
75% R, 25% E	(2.45)	4120	25	3300	12	3700
50% R, 50% E	(2.23)	3750	29	2900	-5	3940
25% R, 75% E	(2.01)	3380	26	2660	-8	3680
Diabase, I	3.00	5050	67	3010	33	3810
Diorite, I	3.10	5220	67	3130	33	3930
Diatomite, S:						
Ditomaceous earth	(0.87)	1470	62	910		
Dolomite, S	2.88	4870	67	2910	43	3400
Earth, loam, S:						
Dry	(1.84)	3030	35	2240	-12	3520
Damp	(2.00)	3370	40	2400	-4	3520
Wet, mud	(1.75)	2940	0	2940	-20	3520
Earth-rock mixtures:						
75% E, 25% R	(2.01)	3380	26	2660	-8	3680
50% E, 50% R	(2.23)	3750	29	2900	-5	3940
25% E, 75% R	(2.45)	4120	25	3300	12	3700
Feldspar, I	2.62	4410	67	2640	33	3320
Felsite, I	2.50	4210	67	2520	33	3170
Fluorite, S	3.10	5220	67	3130		
Gabbro, I	3.10	5220	67	3130	33	3940
Galena, lead ore*	7.51	12630	67	7570		
Gneiss, M	2.71	4550	67	2720	33	3420
Gob, mining refuse	(1.75)	2940	0	2940	-20	3520
Gravel, average graduation, S:						
Dry	(1.79)	3020	15	2610	-7	3240
Wet	(2.09)	3530	5	3350	-3	3640
Granite, I	2.69	4540	72	2640	33	3410
Gumbo, S:						
Dry	(1.91)	3230	50	2150	-10	3570
Wet	(1.99)	3350	67	2020	-10	3720
Gypsum, S	2.43	4080	72	2380		
Hematite, iron ore*	5.08	8560	75	4880		
Hessite, silver ore*	8.50	14300	67	8560		
Ice	0.93	1560	67	930		
Ilmenite, titanium ore*	4.75	8000	69	4730		
Kaolinite, S:						
Dry	(1.91)	3230	50	2150		
Wet	(1.99)	3350	67	2010		

Approximate material characteristics

Material	sp gr	Cubic yards, in cut-weight, lb	Cubic yards, loose		Cubic yards in fill	
			Percent swell	Weight, lb	Swell or shrink, %	Weight, lb
Lignite	(1.25)	2100	65	1270		
Lime						2220
Limestone, S	2.61	4380	63	2690	36	3220
Linnaeite, cobalt ore*	4.89	8230	67	4930		
Limonite, iron ore*	3.80	6400	55	4140		
Loam, earth, S:						
Dry	(1.84)	3030	35	2240	-12	3520
Damp	(2.00)	3370	40	2400	-4	3520
Wet, Mud	(1. 75)	2940	0	2940	-20	3520
Loess, S:						
Dry	(1.91)	3220	35	2380	-10	3570
Wet	(1.99)	3350	40	2400	-10	3720
Magnesite, magnesium ore*	3.00	5050	50	3360		
Magnetite, iron ore*	5.04	8470	54	5520		
Marble, M	2.68	4520	67	2700	33	3400
Marl, S	2.23	3740	67	2240	33	2820
Masonry, rubble	2.33	3920	67	2350	33	2950
Millerite, nickel ore*	5.65	9530	67	5710		
Molybdenite, molybdenum ore*	4.70	7910	67	4750		
Mud, S	(1.75)	2940	0	2940	-20	3520
Muscovite, mica ore*	2.89	4860	67	2910		
Niccolite, nickel ore*	7.49	12600	67	7550		
Orpiment, arsenic ore*	3.51	5900	50	3940		
Pavement:						
Asphalt	1.93	3240	50	1940	0	3240
Brick	2.41	4050	67	2430	33	3050
Concrete	2.35	3960	67	2370	33	2980
Macadam	1.69	2840	67	1700	0	2840
Wood block	0.97	1630	72	950	33	1220
Peat	(0.70)	1180	33	890		
Phosphorite, phosphate rock, S	3.21	5400	50	3600		
Porphyry, I	2.74	4630	67	2770	33	3480
Potash, S	2.20	3700	50	2470		
Pumice, I	0.64	1080	67	650		
Pyrites, iron ore*	5.07	8540	67	5110		
Pyrolusite, manganese ore*	4.50	7560	50	5050		
Quartz, I	2.59	4360	67	2610	33	3280
Quartzite, M	2.68	4520	67	2710	33	3400
Realgar, arsenic ore*	3.51	5900	50	3930		
Rhyolite, I	2.40	4050	67	2420	33	3040
Riprap rock, average	2.67	4500	72	2610	43	3150
Rock-earth mixtures:						
75% R, 25% E	(2.45)	4120	25	3300	12	3700
50% R, 50% E	(2.23)	3750	29	2900	-5	3940
25% R, 75% E	(2.01)	3380	26	2660	-8	3680
Salt, rock, S	2.18	3670	67	2200		

Approximate material characteristics

Material	sp gr	Cubic yards, in cut- weight, lb	Cubic yards, loose		Cubic yards in fill	
			Percent swell	Weight, lb	Swell or shrink, %	Weight, lb
Sand, average graduation, S:						
Dry	(1.71)	2880	11	2590	-11	3240
Wet	(1.84)	3090	5	3230	-11	3460
Sandstone, S	2.42	4070	61	2520	34	3030
Scheelite, tungsten ore*	5.98	10100	67	6050		
Schist, M	2.59	4530	67	2710	33	3410
Serpentine, asbestos ore*	2.62	4440	67	2650		
Shale, S	2.64	4450	50	2970	33	3350
Silt, S	(1.93)	3240	36	2380	-17	3890
Siltstone, S	2.42	4070	61	2520	-11	4560
Slag:						
Furnace	2.87	4840	98	2690	65	2930
Sand	(0.83)	1400	11	1260	-11	1570
Slate, M	2.68	4500	77	2600	33	3380
Smaltite, cobalt ore*	6.48	10970	67	6560		
Snow:						
Dry	(0.13)	220	0	220		
Wet	(0.51)	860	0	860		
Soapstone, talc ore*	2.70	4550	67	2720		
Sodium niter, chile saltpeter	2.20	2710	50	2470		
Stibnite, antimony ore*	4.58	7710	67	4610		
Sulfur	2.00	3450	50	2310		
Syenite, I	2.64	4460	67	2670	33	3350
Taconite, iron ore*	3.18	5370	60	3360		
Talc, M	2.70	4640	67	2780	33	3490
Topsoil, S	(1.44)	2430	56	1620	-26	3280
Trachyte, I	2.40	4050	67	2420	33	3050
Trap rock, igneous rocks, I	2.79	4710	67	2820	33	3540
Trash				400	-50	800
Tuff, S	2.41	4050	50	2700	33	3050
Witherite, barium ore*	4.29	7230	67	4320		
Wolframite, tungsten ore*	7.28	12280	67	7350		
Zinc blende, zinc ore*	4.02	6780	67	4060		
Zincite, zinc ore*	5.68	9550	67	5710		

Key to table:
I—igneous rock. S—sedimentary rock. M—metamorphic rock.
*—ores in the mineral state, with no gangues. Adjust for percentage of mineral bearing gangue or rock to estimate weight of entire ore body, as explained previously in text.
( )—apparent specific gravity, as material is not solid
   Weights per cubic yard in cut are subject to average +10 percent variation. Swell and shrinkage factors for loose condition and embankment are subjectto average +33 percent variation. Weights in loose condition and in embankment are subject to adjustments in accordance with modified swell and shrinkage factors.